edexcel
advancing learning, changing lives

Edexcel AS Chemistry Revision Guide

REVISION GUIDE

David Craggs
Philip Dobson
Geoff Wright

A PEARSON COMPANY

Published by Pearson Education Limited, a company incorporated in England and Wales, having its registered office at Edinburgh Gate, Harlow, Essex, CM20 2JE. Registered company number: 872828

Edexcel is a registered trade mark of Edexcel Limited

First published 2009
10 9 8 7

British Library Cataloguing in Publication Data
A catalogue record for this book is available from the British Library

ISBN 978 1 846905 97 1

External project management by Gillian Lindsey
Edited by Tony Clappison
Typeset by Tech-Set Ltd, Gateshead
Original illustrations © Pearson Education 2009
Illustrated by Tech-Set Ltd, Gateshead and Wearset Ltd
Cover photo © Shutterstock

Printed in Malaysia, (CTP-VVP)

Acknowledgements
Where exam questions are taken from papers specified at the end of the question, these are reproduced by kind permission of Edexcel.

The publishers are grateful to Damian Riddle for writing 'Answering multiple choice and extended questions', Anne Scott and Elizabeth Swinbank at University of York Science Education Group for writing the Revision techniques section and to John Apsey for his collaboration in reviewing this book.

Every effort has been made to contact copyright holders of material reproduced in this book. Any omissions will be rectified in subsequent printings if notice is given to the publishers.

Disclaimer
This material has been published on behalf of Edexcel and offers high-quality support for the delivery of Edexcel qualifications.

This does not mean that the material is essential to achieve any Edexcel qualification, nor does it mean that it is the only suitable material available to support any Edexcel qualification. Edexcel material will not be used verbatim in setting any Edexcel examination or assessment. Any resource lists produced by Edexcel shall include this and other appropriate resources.

Copies of official specifications for all Edexcel qualifications may be found on the Edexcel website - www.edexcel.com

Contents

How to use this Revision Guide

Welcome to your **Edexcel AS Chemistry Revision Guide**.

This unique guide provides you with tailored support, written by Senior Examiners. They draw on real 'ResultsPlus' exam data from past A-level exams, and have used this to identify common pitfalls that have caught out other students, and areas on which to focus your revision. As you work your way through the topics, look out for the following features throughout the text:

ResultsPlus Examiner Tip

These sections help you perform to your best in the exams by highlighting key terms and information, analysing the questions you may be asked, and showing how to approach answering them. All of this is based on data from real-life A-level students!

ResultsPlus Watch Out!

The examiners have looked back at data from previous exams to find the common pitfalls and mistakes made by students – and guide you on how to avoid repeating them in your exam.

Quick Questions

Use these questions as a quick recap to test your knowledge as you progress.

Thinking Task

These sections provide further research or analysis tasks to develop your understanding and help you revise.

Worked Examples

The examiners provide step-by-step guidance on complex equations and concepts.

Each topic also ends with:

Topic Checklist

This summarises what you should know for this topic, which specification point each checkpoint covers and where in the guide you can revise it. Use it to record your progress as you revise.

ResultsPlus Build Better Answers

Here you will find sample exam questions with exemplar answers, examiner tips and a commentary so you can see how to get the highest marks.

Practice Exam Questions

Exam-style questions, including multiple-choice, offer plenty of practice ahead of the written exams.

Both Unit 1 and Unit 2 conclude with a **Practice Unit Test** to test your learning. These are not intended as timed, full-length papers, but provide a range of exam-style practice questions covering the range of content likely to be encountered within the unit test.

The final Unit consists of advice and support on the practical exercises which assess your chemistry laboratory skills, giving guidance to help you make inferences from qualitative observations, carry out quantitative measurements and prepare a named substance.

Answers to all the in-text questions, as well as detailed, mark-by-mark answers to the practice exam questions, can be found at the back of the book.

We hope you find this guide invaluable. Best of luck!

Getting started can be the hardest part of revision, but don't leave it too late. Revise little and often! Don't spend too long on any one section, but revisit it several times, and if there is something you don't understand, ask your teacher for help. Just reading through your notes is not enough. Take an active approach using some of the revision techniques suggested below.

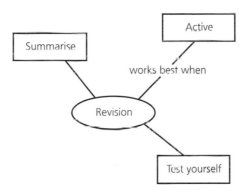

Summarising key ideas

Make sure you don't end up just copying out your notes in full. Use some of these techniques to produce condensed notes.
- Tables and lists to present information concisely.
- Index cards to record the most important points for each section.
- Flow charts to identify steps in a process.
- Diagrams to present information visually.
- Spider diagrams, mind maps and concept maps to show the links between ideas.
- Mnemonics to help you remember lists.
- Glossaries to make sure you know clear definitions of key terms

Include page references to your notes or textbook. Use colour and highlighting to pick out key terms.

Active techniques

Using a variety of approaches will prevent your revision becoming boring and will make more of the ideas stick. Here are some methods to try.
- Explain ideas to a partner and ask each other questions.
- Make a podcast and play it back to yourself.
- Use PowerPoint to make interactive notes and tests.
- Search the Internet for animations, tests and tutorials that you can use.
- Work in a group to create and use games and quizzes.

If you use resources from elsewhere, make sure they cover the right content at the right level.

Test yourself

Once you have revised a topic, you need to check that you can remember and apply what you have learnt.
- Use the questions from your textbook and this revision guide.
- Get someone to test you on key points.
- Try some past exam questions.

Answering extended and multiple-choice questions

Section A of Unit tests 1 and 2 contain objective test (multiple-choice) questions. Section B contains a mixture of short-answer and extended-answer questions, including the analysis, interpretation and evaluation of experimental and investigative activities. Section C (unit 2 only) includes contemporary context questions. In all sections you may be required to apply your knowledge and understanding of chemistry to situations that you have not seen before.

Multiple-choice questions

For each question there are four possible answers, labelled A, B, C and D. A good multiple-choice question (from an examiner's point of view) gives the correct answer and three other possible answers, which all seem plausible.

The best way to answer a multiple-choice question is to read the question and try to answer it *before* looking at the possible answers. You may need to do some calculations – space is provided on the question paper for rough working. If the answer you thought of or calculate is among the possible answers – job done! Just have a look at the other possibilities to convince yourself that you were right.

If the answer you thought of isn't there, look at the possible answers and try to eliminate wrong answers until you are left with the correct one.

You don't lose any marks by having a guess (if you can't work out the answer) – but you won't score anything by leaving the answer blank. If you narrow down the number of possible answers, the chances of having a lucky guess at the right answer will increase.

To indicate the correct answer, put a cross in the box following the correct statement. If you change your mind, put a line through the box and fill in your new answer with a cross.

How Science Works

The idea behind 'How Science Works' is to give you insight into the ways in which scientists work: how an experiment is designed, how theories and models are put together, how data is analysed, how scientists respond to factors such as ethics and so on.

Many of the HSW criteria require practical or investigative skills and will be tested as part of your assessed practical work. However, there will be questions on the written units that cover all the HSW criteria. Some of these questions will involve data or graph interpretation, including determining quantities (with appropriate units) from the gradient and intercept of a graph.

Another common type of HSW question will be on evaluating various steps in an experiment. For example,
- explain or justify why a particular piece of apparatus is used
- identify possible sources of systematic or random error
- explain why we use an instrument in a particular way
- what safety precautions would be relevant, and why?

You may be asked questions involving designing an investigation: these are likely to involve pieces of familiar practical work.

Other HSW questions may concentrate on issues surrounding the applications and implications of science (including ethical issues), or on using a scientific model to make predictions.

Extended questions

Remember that if part of a question is worth 6 marks, you need to make six creditworthy points. Think about the points that you will make and put them together in a logical sequence when you write your answer. On longer questions, the examiners will be looking at your QWC (Quality of Written Communication) as well as the answer you give.

Chemical quantities and formulae

Atoms, elements and compounds

- An **element** is a substance that cannot be broken down by chemical means into other substances.
- All the atoms in an element have the same **atomic number**, **Z**, which is the number of protons in an atom.
- An **atom** is the smallest part of an element that can take part in a chemical change. Atoms have no charge, they are neutral.
- The **mass number**, **A**, of an atom is the sum of protons and neutrons in that atom.
- **Isotopes** are atoms of the same element (same number of protons) with different numbers of neutrons, therefore they have different mass numbers.

This leads to an important difference between mass number and **relative atomic mass**, A_r.

The **relative atomic mass** of an element is the average (weighted) mass of the element's isotopes relative to one-twelfth of the mass of a carbon-12 atom.

Thinking Task

Will the reactions of ^{24}Mg and ^{23}Mg with oxygen be any different? If so, how? If not, why?

ResultsPlus
Examiner tip

Remember to check that your answer to any calculation is reasonable. In this case, ask does it lie somewhere between 46 and 50?

Worked Example

The element titanium has the following isotopic abundances:

Isotope	^{46}Ti	^{47}Ti	^{48}Ti	^{49}Ti	^{50}Ti
Relative abundance/%	8.0	7.3	73.8	5.5	5.4

So its relative atomic mass is worked out as follows:

$$A_r = \frac{(46 \times 8.0) + (47 \times 7.3) + (48 \times 73.8) + (49 \times 5.5) + (50 \times 5.4)}{100}$$

$$= 47.9$$

- A **compound** is formed when two or more elements are chemically bonded together.
- A **molecule** is the smallest part of a *covalent* compound or element that can exist on its own.
- *Ionic* compounds are formed from ions.
- An **ion** is formed when an atom gains or loses one or more electrons – ions are, therefore, charged particles.

ResultsPlus
Watch out!

Ionic compounds are *not* made of molecules or atoms – they are *ionic*! So don't even mention molecules or atoms when answering questions about ionic compounds, ionic lattices or electrolysis.

The **relative molecular mass**, M_r, of a compound is the sum of the relative atomic masses of all its atoms. This is also called **relative formula mass**.

Calculating amounts of chemicals

The **mole** is the unit of chemical amount, or **amount of substance**.
- 1 mole is the amount of substance that contains as many particles as there are atoms in exactly 12 g of carbon-12.
- The number of atoms in exactly 12 g of carbon-12 is 6.02×10^{23}. This is known as the **Avogadro constant (L)** and has the unit particles per mole (in calculations, mol^{-1}).
- The mass, in grams, of 1 mole of substance is known as the **molar mass**, M_r.
- The molar mass of an element is its relative atomic mass in grams.
- The molar mass of molecules of an element or compound is the relative molecular mass in grams.
- The amount of substance (the number of moles) $= \dfrac{\text{mass of substance}}{\text{molar mass}}$

- The amount of substance in a solution $=$ volume of solution $\times$ concentration of solution.

ResultsPlus
Examiner tip

You should be able to calculate molar mass of a compound from the atomic masses given in the Periodic Table on the back of your exam paper.

Concentrations can also be expressed in **parts per million, ppm**. This could be the amount by volume or by mass.

$$ppm = \frac{\text{number of parts of the chemical}}{\text{number of parts it is contained in}} \times 10^6$$

Empirical and molecular formulae

The **empirical formula** of a compound shows the simplest atom ratio of the elements in that compound. Different compounds may have the same empirical formula – for example, ethyne (C_2H_2) and benzene (C_6H_6) have the empirical formula CH.

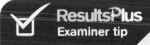

Worked Example

What is the empirical formula of a compound that contains 40.00% carbon, 6.67% hydrogen and 53.33% oxygen by mass?

Step	What to do			
1	Element symbol	C	H	O
2	Mass or percentage	40.00	6.67	53.33
3	Divide by A_r	$\dfrac{40.00}{12.0}$	$\dfrac{6.67}{1.0}$	$\dfrac{53.33}{16.0}$
		3.33	6.67	3.33
4	Divide by smallest number	$\dfrac{3.33}{3.33}$	$\dfrac{6.67}{3.33}$	$\dfrac{3.33}{3.33}$
5	Smallest whole ratio	1	2	1
6	Write the formula	CH_2O		

ResultsPlus
Examiner tip

The ratios might not be whole numbers – for example, if the ratio C : H : O was 1.48 : 2.96 : 1 then you must double each value to get 3 : 6 : 2; so the empirical formula is $C_3H_6O_2$. This shows that you must not round off the numbers too early during the calculation. Rounding too early may give an incorrect formula.

You use these same steps for calculating empirical formulae whether the given data is percentage mass of elements in the compound or the actual masses of elements in a sample.

The **molecular formula** of a covalent or ionic compound shows how many atoms/ ions of each element combine to make that compound. It is an exact multiple of the empirical formula.

Worked Example

For the compound whose empirical formula is CH_2O, the molecular mass was found to be 180. What is its molecular formula?

$$\text{Molecular formula} = \text{empirical formula} \times \frac{\text{molecular mass}}{\text{empirical formula mass}}$$

C = 12, H = 1, O = 16, so the empirical formula mass is 12 + 2 + 16 = 30

So the molecular formula = $CH_2O \times \dfrac{180}{30} = C_6H_{12}O_6$

Quick Questions

1 **a** Calculate the number of moles and the number of atoms in:
      i 22.0 g of carbon dioxide,   **ii** 1.000 kg of water.
   b Calculate the number of moles of potassium ions in 5.6 dm³ of blood where the concentration is 4.1 mmol dm⁻³.
2 Use the following isotopic abundance data for silicon to calculate its relative atomic mass.

Isotope	^{28}Si	^{29}Si	^{30}Si
Relative abundance/%	92.2	4.7	3.1

3 **a** What is the empirical formula of a compound containing 30.87% Na, 47.65% Cl and 21.48% O?
   b A sample of a compound was found to have a molecular mass of 267 and to contain 16.2 g Al and 63.9 g Cl.
      i What is its empirical formula?   **ii** What is its molecular formula?

ResultsPlus
Watch out!

Use the same number of significant figures in your calculation and answer as there are in the data you are given. Marks may be lost for giving an inappropriate number of significant figures.

Chemical equations and reacting masses

Balanced full equations

A balanced chemical equation tells us not only what is reacting and what is produced, but also in what proportions the atoms combine.

Worked Example

For example, the reaction between hydrogen and oxygen to produce water (steam) is described in words as:

$$\text{hydrogen} + \text{oxygen} \rightarrow \text{water}$$

If we just write the formulae:

$$H_2 + O_2 \rightarrow H_2O$$

the equation is not **balanced** – there are not the same numbers of oxygen atoms on each side. So H_2O has to be doubled to give:

$$H_2 + O_2 \rightarrow 2H_2O$$

But now we have 2 H atoms on the left-hand side and 4 on the other. We correct this by putting $2H_2$ on the left:

$$2H_2 + O_2 \rightarrow 2H_2O$$

This is the **balanced chemical equation**. A balanced chemical equation has the same number of the same atoms on each side – they are just arranged differently.

State symbols

These are important and should be included in all chemical equations as a matter of habit:
- **(s)** for solids
- **(l)** for liquids
- **(g)** for gases
- **(aq)** for aqueous solutions.

So for the above reaction we should write:

$$2H_2(g) + O_2(g) \rightarrow 2H_2O(g)$$

Ionic equations

Ions that appear on both sides of an equation *in the same state* are **spectator ions**. They take no part in the reaction and can be removed from the full equation to give an **ionic equation**.

Worked Example

Write the ionic equation for the reaction when an iron nail in copper sulfate solution displaces the copper:

$$Fe + CuSO_4 \rightarrow Cu + FeSO_4$$

Putting in the state symbols and showing all the separate ions gives:

$$Fe(s) + Cu^{2+}(aq) + SO_4^{2-}(aq) \rightarrow Cu(s) + Fe^{2+}(aq) + SO_4^{2-}(aq)$$

Ions *in the same state* on both sides are identified:

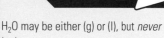

$$Fe(s) + Cu^{2+}(aq) + \cancel{SO_4^{2-}(aq)} \rightarrow Cu(s) + Fe^{2+}(aq) + \cancel{SO_4^{2-}(aq)}$$

The ionic equation is therefore:

$$Fe(s) + Cu^{2+}(aq) \rightarrow Cu(s) + Fe^{2+}(aq)$$

Reacting masses

Balanced chemical equations are used to work out how much reactant is needed to make a certain amount of product.

Worked Example

How many tonnes of limestone (calcium carbonate) have to be heated to produce 100 tonnes of quicklime (calcium oxide)?

Step	What to do	
1	Balanced equation	$CaCO_3(s) \rightarrow CaO(s) + CO_2(g)$
2	Amounts involved	1 mol $\rightarrow$ 1 mol (ignore CO_2)
3	Molar masses	$100\,g \rightarrow 56.0\,g$
4	Divide both sides by 56	$\dfrac{100\,g}{56} \rightarrow \dfrac{56.0\,g}{56}$
		$1.79\,g \rightarrow 1.0\,g$
		$1.79\,t \rightarrow 1.0\,t$
5	Multiply both sides by 100	$179\,t \rightarrow 100\,t$

179 tonnes of limestone have to be heated.

Finding chemical equations from experimental data

By measuring the masses of reactants and products, it is possible to confirm the balanced equation for the reaction.

Worked Example

When 11.2 g of iron reacted with excess chlorine, 32.0 g of iron(III) chloride were formed. What is the equation for the reaction?

Amount of iron $= \dfrac{11.2\,g}{56\,g\,mol^{-1}} = 0.200\,mol$

Amount of iron(III) chloride $= \dfrac{32.0\,g}{162.5\,g\,mol^{-1}} = 0.197\,mol$

The limitation of the technique and accuracy of measurement leads to the conclusion that 2 moles of iron in excess chlorine produces 2 moles of iron(III) chloride. So the equation must be:

$$2Fe + 3Cl_2 \rightarrow 2FeCl_3$$

Experimental data for reacting masses can also be used to find the empirical formula of a compound (see the method on the previous page).

Quick Questions

1. The following equations are not balanced. Copy and balance them.
   a $KBrO_3(s) + C(s) \rightarrow KBr(s) + CO(g)$
   b $ZnS(s) + O_2(g) \rightarrow ZnO(s) + SO_2(g)$
2. Write the full and ionic equations for the following reactions:
   a barium chloride solution reacting with sodium sulfate solution
   b nitric acid reacting with calcium hydroxide solution
3. What is the equation for the reaction when 0.654 g of zinc powder react with excess silver nitrate solution to produce 2.16 g silver?

Reactions with gases

- The **molar volume** of a gas is the volume of 1 mole of it.
- For all gases, this is $24\,dm^3\,mol^{-1}$ at **standard temperature (25°C or 298 K)** and **pressure (1 atm)**.

We use chemical equations to calculate volumes of gases in the same way as reacting masses.

Worked Example

What volume of oxygen at standard temperature and pressure is needed to burn exactly $6\,dm^3$ methane according to the following equation?

$$CH_4(g) + 2O_2(g) \rightarrow CO_2(g) + 2H_2O(g)$$

All gases have the same molar volume at s.t.p. so we can use the molar ratios in the balanced equation as the proportionate reacting volumes.

$$6\,dm^3 \text{ of methane} = \tfrac{6}{24} \text{ molar volume} = \tfrac{1}{4} \text{ molar volume}.$$

From the equation:

1 molar volume of methane reacts with 2 molar volumes of oxygen

$\tfrac{1}{4}$ molar volume of methane reacts with $\tfrac{1}{2}$ molar volume of oxygen

$$\tfrac{1}{2} \text{ molar volume of oxygen} = \tfrac{1}{2} \times 24\,dm^3$$
$$= 12\,dm^3 \text{ of oxygen.}$$

Worked Example

What volume of carbon dioxide (at s.t.p.) is formed when 23 g of hexane is burned in excess oxygen?

Write the balanced equation, assuming complete combustion:

$$C_6H_{14}(l) + 9\tfrac{1}{2}O_2(g) \rightarrow 6CO_2(g) + 7H_2O(g)$$

Molar mass of $C_6H_{14} = (6 \times 12) + (14 \times 1) = 86\,g\,mol^{-1}$

Therefore $23\,g\ C_6H_{14} = \dfrac{23\,g}{86\,g\,mol^{-1}} = 0.27\,mol$

From the equation:

1 mol of C_6H_{14} produces 6 mol of CO_2

0.27 mol of C_6H_{14} produces $0.27\,mol \times 6 = 1.60\,mol$ of CO_2

This amount of $CO_2 = 1.60\,mol \times 24\,dm^3\,mol^{-1} = 38.5\,dm^3$ at s.t.p.

or $39\,dm^3$ to 2 significant figures.

Finding chemical equations from experimental data

Ammonia gas can react with oxygen to produce nitrogen and water. Excess ammonia was reacted with 4.5 dm³ of oxygen (at s.t.p.) and 3.0 dm³ of nitrogen were collected. What is the equation for this reaction?

The volumes of gases at s.t.p. are proportional to their amounts, therefore:

$$4.5 \text{ mol oxygen} \rightarrow 3 \text{ mol nitrogen}$$

Hence, (dividing by 3):

$$1.5 \text{ mol oxygen} \rightarrow 1 \text{ mol nitrogen}$$

Or, more usefully:

$$3 O_2 \rightarrow 2 N_2$$

$2N_2$ on the product side means that there must be $4NH_3$ reacting and $3O_2$ on the reactant side will produce $6H_2O$.

Therefore the equation must be:

$$4NH_3 + 3O_2 \rightarrow 2N_2 + 6H_2O$$

Quick Questions

1   a   Calculate what volume of oxygen will react exactly with 18.0 dm³ of ammonia at s.t.p. according to the equation:
$$4NH_3 + 3O_2 \rightarrow 2N_2 + 6H_2O$$

   b   Calculate what volume of steam will be produced when 500 cm³ of hydrogen sulfide react with excess oxygen according to the equation:
$$2H_2S(g) + 3O_2(g) \rightarrow 2SO_2(g) + 2H_2O(g)$$

2   Calculate what volume of carbon dioxide is produced at s.t.p. when a marble chip weighing 0.500 g is completely decomposed by heating according to the equation:
$$CaCO_3(s) \rightarrow CaO(s) + CO_2(g)$$

3   What is the equation for the reaction when 750 cm³ of ammonia react with excess oxygen to produce 750 cm³ of nitrogen monoxide (NO(g)) and steam?

Thinking Task

Why is temperature an important consideration in calculations involving gases? What is the relationship between degrees Celsius and Kelvin? Why is the Kelvin scale used in thermodynamics?

Percentage yield and atom economy

Percentage yield

Remember – no atoms are gained or lost in a chemical reaction. However, it is unusual to obtain the theoretical maximum amount of a product:

- the reaction may be reversible
- some of the reactants or product may be left behind in the apparatus, such as when filtering or pouring
- some of the reactants may react in ways different to the expected reaction (**side reactions**, which produce **by-products**).

The amount of a product obtained is the **yield**. When compared with the maximum theoretical amount as a percentage, it is called the **percentage yield**:

$$\text{percentage yield} = \frac{\text{actual mass of product}}{\text{theoretical mass of product}} \times 100\%$$

For example, if a chemical reaction actually makes 12 g of product, when in theory it should produce 60 g then:

$$\text{percentage yield} = \frac{12\,g}{60\,g} \times 100\%$$
$$= 20\%$$

ResultsPlus
Examiner tip

In the written and practical tests, you will be expected to be able to calculate the percentage yield when preparing pure crystals of a salt or double salt from an aqueous solution, and suggest reasons for a low yield.

Worked Example

Magnesium reacts with oxygen according to the equation:
$$2Mg(s) + O_2(g) \rightarrow 2MgO(s)$$
When a student heated a 2.00 g coil of magnesium ribbon in a crucible, she produced 3.00 g of the oxide. What is the theoretical yield? And what percentage yield was achieved?

Calculating the theoretical yield is a reacting masses calculation, see page 9.

Step	What to do		
1	Balanced equation	$2Mg(s) + O_2(g) \rightarrow 2MgO(s)$	
2	Amounts involved	2 mol	$\rightarrow$ 2 mol (ignore O_2)
3	Molar masses	48.6 g	$\rightarrow$ 80.6 g
4	Divide both sides by 48.6	$\dfrac{48.6\,g}{48.6}$	$\rightarrow \dfrac{80.6\,g}{48.6}$
		1.00 g	$\rightarrow$ 1.66 g
5	Multiply both sides by 2.00	2.00 g	$\rightarrow$ 3.32 g

So theoretical yield from 2.00 g of magnesium is 3.32 g of magnesium oxide.

$$\text{Percentage yield} = \frac{\text{actual mass of product}}{\text{theoretical mass of product}} \times 100\%$$
$$= \frac{3.00\,g}{3.32\,g} \times 100\%$$
$$= 90.4\%$$

ResultsPlus
Watch out!

When you prepare a soluble salt it is important not to evaporate the solution to dryness by heating it. The crystals may decompose or lose the water of crystallization and you will not get as much as you ought.

Yields in salt preparation

Salts are ionic compounds. The positive ion, or cation, is usually a metal ion or an ammonium ion, NH_4^+. The negative ion, or anion, comes from an acid such as SO_4^{2-} in H_2SO_4.

A **double salt** contains more than one cation or anion. They form when a solution of two simple salts crystallizes to form a single substance.

- Salts can be produced by neutralizing acids with an alkali, carbonate or metal oxide or hydroxide.

- Salts can be produced by reacting acids with reactive metals.
- A soluble salt must be crystallized from a saturated solution – concentrate the solution by driving off some of the water, leaving to evaporate and then filtering.
- An insoluble salt forms a precipitate and can be filtered off, washed and dried.

Worked Example

A student prepared a sample of copper(II) nitrate crystals by neutralizing excess dilute nitric acid with 12.35 g copper(II) carbonate, then filtering and evaporating the solution. The crystals were dehydrated and had a mass of 18.05 g. Calculate the percentage yield.

Again, this is a reacting masses calculation, so 12.35 g $CuCO_3$ produces a theoretical yield of 18.75 g $Cu(NO_3)_2$.

But only 18.05 g $Cu(NO_3)_2$ was produced.

$$\text{Percentage yield} = \frac{18.05\,g}{18.75\,g} \times 100\%$$

$$= 96.3\%$$

Atom economy

Atom economy is a measure of the amount of starting materials that end up as useful products – it is *not* the same as the **yield**. Atom economy is calculated using the balanced chemical equation for a reaction, assuming it produces 100% yield:

$$\text{atom economy} = \frac{\text{mass of atoms in required product}}{\text{total mass of atoms in reactants}} \times 100\%$$

- A reaction with a high atom economy makes use of most of the atoms of the reactants, with few wasted as by-products.
- This reduces the amount of waste products a company has to deal with, which in turn reduces the cost of waste treatment.
- Atom economy can be improved by finding uses for any by-products.

Worked Example

Calculate the atom economy of using copper oxide and sulfuric acid to make copper sulfate in solution:

$$CuO(s) + H_2SO_4(aq) \rightarrow CuSO_4(aq) + H_2O(l)$$

$$\text{Atom economy} = \frac{\text{mass of copper sulfate}}{\text{masses of copper oxide + sulfuric acid}} \times 100\%$$

$$= \frac{63.5 + 32.0 + (4 \times 16.0)}{(63.5 + 16.0) + (2 + 32.0 + (4 \times 16.0))} \times 100\%$$

$$= 89.9\%$$

ResultsPlus
Watch out!

Percentage yield and atom economy are different. Percentage yield is about the efficiency of the process of converting reactants into products. Atom economy is about the proportion of reactant atoms that have been converted into the desired product.

Quick Questions

1 A student prepared a sample of the fertilizer ammonium sulfate by adding ammonia solution to sulfuric acid:

$$2NH_3(aq) + H_2SO_4(aq) \rightarrow (NH_4)_2SO_4(aq)$$

   a Calculate the theoretical maximum yield that can be obtained by reacting 25.0 cm³ of 2.0 mol dm⁻³ ammonia solution with an excess of sulfuric acid.
   b If the actual mass obtained was 3.12 g, calculate the percentage yield and suggest reasons why the yield is less than 100%.
2 Calculate the atom economy to make copper nitrate in the reaction:

$$CuCO_3(s) + 2HNO_3(aq) \rightarrow Cu(NO_3)_2(aq) + CO_2(g) + H_2O(l)$$

Topic 1: Formulae, equations and amounts of substance checklist

By the end of this topic you should be able to:

Revision spread	Checkpoints	Specification section	Revised	Practice exam questions
Chemical quantities and formulae	Understand the terms atom, element, ion, molecule and compound	1.3a	☐	☐
	Understand the terms relative atomic mass, amount of substance, molar mass and parts per million	1.3c	☐	☐
	Calculate the amount of substance in a solution of known concentration	1.3d	☐	☐
	Understand and be able to perform calculations using the Avogadro constant	1.3h	☐	☐
Chemical equations and reacting masses	Write balanced equations (full and ionic), including state symbols	1.3b	☐	☐
	Understand the terms empirical and molecular formulae	1.3a	☐	☐
	Use chemical equations to calculate reacting masses, or masses of reactants and products to find the equation for a reaction	1.3e	☐	☐
	Interpret the results of simple chemical tests (see also Unit 3); relate the observations of test-tube reactions to the state symbols used in equations, and write full and ionic equations for these reactions	1.3k	☐	☐
Reactions with gases	Use chemical equations to calculate volumes of gases or masses of reacting gases to find the equation	1.3f	☐	☐
Percentage yield and atom economy	Use chemical equations and experimental results to calculate percentage yields and atom economies; explain the advantages of a high atom economy	1.3g	☐	☐
	Analyse and evaluate the results obtained from finding a formula or confirming an equation by experiment	1.3i	☐	☐
	Make a salt and calculate the percentage yield of product	1.3j	☐	☐

ResultsPlus
Build Better Answers

1 In an experiment to prepare 1-bromobutane, 4.86 g of butan-1-ol reacted with concentrated sulfuric acid and sodium bromide. The equation for the reaction is:

$$C_4H_9OH(l) + NaBr(aq) + H_2SO_4(aq) \rightarrow C_4H_9Br(l) + H_2O(l) + NaHSO_4(aq)$$

After purification, 5.02 g of 1-bromobutane was collected.

a Calculate the amount, in moles, of butan-1-ol used in the preparation.   (2)

✓ Examiner tip

First work out the M_r of $C_4H_9OH = (4 \times 12) + (9 \times 1) + 16 + 1 = 74$ (1)

Mass = relative formula mass × amount, hence:

$$amount = \frac{mass}{relative\ formula\ mass}$$

$$= \frac{4.86\ g}{74\ g\ mol^{-1}}$$

$$= 0.066\ mol \quad \text{(1 for correct answer showing working)}$$

Sometimes in this sort of calculation you will be *given* the relative molecular mass of the product. Read the question carefully – you may lose a mark in the percentage yield calculation if you use the wrong molar mass!

b i Calculate the mass of 1-bromobutane that would be formed (theoretical yield) if all the butan-1-ol was converted into 1-bromobutane. (2)

> ☑ **Examiner tip**
>
> From the equation, 1 mol butan-1-ol produces 1 mol 1-bromobutane. Therefore 0.066 mol butan-1-ol produces 0.066 mol 1-bromobutane. (1)
>
> $$M_r \ C_4H_9Br = 137$$
> $$\text{Mass of 1-bromobutane} = 0.066 \ mol \times 137 \ g \ mol^{-1}$$
> $$= 9.04 \ g \quad (1)$$

ii Calculate the percentage yield. (1)

> ☑ **Examiner tip**
>
> $$\text{Percentage yield} = \frac{\text{actual yield}}{\text{theoretical yield}} \times 100\%$$
> $$= \frac{5.02 \ g}{9.04 \ g} \times 100\%$$
> $$= 55.5\%$$

Practice exam questions

1 a What does the term *empirical formula* mean? (2)

b A compound was found to contain 29.1% sodium, 40.5% sulfur and 30.4% oxygen. Calculate its empirical formula. (3)

2 a What mass of sodium hydrogencarbonate must be heated to produce 100 g of sodium carbonate? The reaction is: (3)
$$2NaHCO_3(s) \rightarrow Na_2CO_3(s) + H_2O(g) + CO_2(g)$$

b What volume of carbon dioxide (at s.t.p.) would be produced at the same time? (1 mole of a gas at s.t.p. occupies 24 dm³) (2)

3 a Aluminium is extracted by electrolysis of Al_2O_3. What is the maximum mass of the metal that could be obtained from 1 tonne of the oxide? (2)

b If the yield of the electrolysis plant is 90%, what mass of the oxide must be used to produce 1 tonne of aluminium? (2)

4 a A solution of hydrated ethanedioic acid ($H_2C_2O_4.2H_2O$) was made by dissolving the solid in water and making it up to 250.0 cm³. If the concentration was 0.0470 mol dm⁻³, what mass of solid was used? (3)

b What mass of anhydrous sodium carbonate (Na_2CO_3) must be used to prepare 250.0 cm³ of a solution of concentration 0.05 mol dm⁻³? (3)

5 A compound is 86% carbon and 14% hydrogen by mass. What is the empirical formula for this compound?

   A CH   **B** CH_2 **C** CH_3 **D** CH_4 (1)

6 At standard temperature and pressure, 32.0 dm³ of O_2 contain the same number of molecules as:

   A 22.4 dm³ of Cl_2 **B** 28.0 dm³ of N_2

   C 32.0 dm³ of H_2 **D** 44.8 dm³ of HCl (1)

7 Given the reaction $CH_4(g) + 2O_2(g) \rightarrow CO_2(g) + 2H_2O(g)$, what mass of oxygen is needed to completely react with 56 g CH_4?

   A 56 g   **B** 112 g   **C** 179.3 g   **D** 224 g (1)

Enthalpy changes and enthalpy level diagrams

When a reaction occurs at *constant pressure* and gives out or takes in energy there is an **enthalpy change**, ΔH, between the system (reactants and products) and its surroundings:

$$\Delta H = H_{products} - H_{reactants}$$

Enthalpy level diagrams show the relative energy levels of reactants and products.

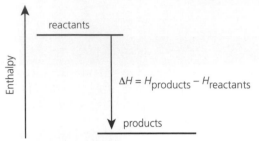

Simple enthalpy level diagram for an exothermic reaction

An **exothermic** reaction has a negative ΔH: $\Delta H < 0$. This is because the products are at a *lower* energy level than the reactants – the products have *less* energy than the reactants. Energy has been *given out* by the system to the surroundings – the surroundings get hotter.

A Bunsen burner flame produces heat energy. This is an exothermic reaction – heat is being given out to the surroundings. This heat comes from the energy content of the methane and oxygen.

The carbon dioxide and water are at a lower energy level:

$$CH_4(g) + 2O_2(g) \rightarrow CO_2(g) + 2H_2O(l) \qquad \Delta H \; -ve$$

An **endothermic** reaction has a *positive* ΔH: $\Delta H > 0$. This is because the products are at a *higher* energy level than the reactants – the products have *more* energy than the reactants. Energy has been *absorbed* by the system from the surroundings – the surroundings get colder.

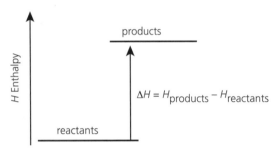

Simple enthalpy level diagram for an endothermic reaction

Melting is an **endothermic** process – heat is absorbed from the surroundings. This heat comes from the energy content of the surroundings, so when an ice cube melts in water the water gets colder.

The melted substance is at a *higher* energy level:

$$H_2O(s) \rightarrow H_2O(l) \qquad \Delta H \; +ve$$

Thinking Task

How can boiling a used hand warmer in water recharge it?

? Quick Questions

1. Draw and label a simple enthalpy level diagram for an ice–water system in which the ice is melting.
2. Some hand warmers and other hot packs use a supersaturated solution of sodium ethanoate to produce heat as it crystallizes. Draw and label a simple enthalpy level diagram for this type of warmer.
3. Solid ammonium nitrate and water are used in cold packs. When 60 g of ammonium nitrate reacted with 200 g water, the energy change was 15.8 kJ. Draw and label a simple enthalpy level diagram for this reaction.
4. Why does an exothermic reaction have a negative value for ΔH?

Measuring enthalpy changes

Standard conditions

Enthalpy changes depend on the temperature and pressure at which the reactions occur. Therefore, a set of standard conditions is necessary to make measurements comparable. Standard conditions are:

- 1 atm pressure at 298 K (25°C)
- Reacting substances are in their normal physical states at 1 atm and 298 K
- Solutions have a concentration of 1 mol dm^{-3}.

An enthalpy change measured under these standard conditions is called a **standard enthalpy change** and has the symbol $\Delta H^{\ominus}$.

$\Delta H^{\ominus}$ is expressed per mole of substance reacting, and the units are kJ mol^{-1}.

Definitions and symbols

- The **standard enthalpy change of reaction**, $\Delta H^{\ominus}$, is the enthalpy change under standard conditions for a reaction in the molar quantities written in the chemical equation.
- The **standard enthalpy change of formation of a compound**, $\Delta H_f^{\ominus}$, is the enthalpy change when 1 mole of the compound is formed from its elements under standard conditions. The standard enthalpy change of formation of any element, in its standard state, is therefore zero. For example, $\Delta H_f^{\ominus} [Cl_2(g)] = 0$.
- The **standard enthalpy change of combustion**, $\Delta H_c^{\ominus}$, is the enthalpy change when 1 mole of the substance is completely burnt in oxygen under standard conditions.
- The **standard enthalpy change of atomization**, $\Delta H_{at}^{\ominus}$, is the enthalpy change when 1 mole of atoms in the gaseous state is formed from the element under standard conditions. For example, $\Delta H_{at}^{\ominus} \left[\frac{1}{2} Cl_2(g) \rightarrow Cl(g)\right]$ is not zero, because energy must be transferred to break bonds in the Cl_2 molecules.
- The **standard enthalpy change of neutralization** is the enthalpy change when 1 mole of water is formed by the reaction of an acid with an alkali under standard conditions.

Enthalpy change from experiments

The energy transferred in a reaction can be calculated from the temperature change of the system (if in solution) or a measured mass of water (in **calorimetry** experiments).
Energy change in surroundings $= m \times c \times \Delta T$

- m is the mass
- c is the **specific heat capacity** (for water, c is 4.2 J g^{-1} K^{-1})
- ΔT is the change in temperature.

This energy change is equal to the enthalpy change for reactions at constant pressure (open to the surroundings). The standard enthalpy change of combustion of a fuel such as ethanol can be determined by burning a known amount of the fuel.

Worked Example

Using a spirit burner and copper can calorimeter, a student heated 200 g water from 22°C to 52°C by burning 1.8 g ethanol. Calculate the enthalpy change of combustion of ethanol.

Energy transferred to the water = mass $\times$ specific heat capacity $\times$ temperature change

$= 200 \text{ g} \times 4.2 \text{ J g}^{-1} \text{K}^{-1} \times (52 - 22) \text{ K}$

$= 25.2 \text{ kJ}$

Energy produced by 1.8 g of ethanol = 25.2 kJ

ResultsPlus
Watch out!

Always give the sign of ΔH even if it is positive – there might be a mark for the correct sign.

ResultsPlus
Examiner tip ✓

You should also be able to evaluate the results obtained from enthalpy change experiments, and comment on sources of error and the assumptions made. It is examined in the Unit 1 paper, and also in Unit 3 Activity d.

Molar mass of ethanol = 46 g mol^{-1}

Energy produced per mole ethanol = 25.2 kJ/1.8 g × 46 g mol^{-1}
$$= 644 \, kJ \, mol^{-1}$$

This is an exothermic reaction so the enthalpy change is = −644 kJ mol^{-1}.

The value of ΔH_c calculated above is much lower than the accepted (data book) value of −1371 kJ mol^{-1}.

A number of assumptions have been made in the calculation:
- all the energy produced by the burning ethanol is transferred to the water
- no heat is absorbed by the calorimeter
- there is no heat loss to the room or from the surface of the water
- the fuel has combusted completely – partial combustion lowers the amount of energy released.

These assumptions give rise to **systematic errors** – inaccuracies in the calculated enthalpy change. **Random errors** also arise from measurement uncertainties – for example in the mass of water or the recorded temperatures.

ResultsPlus
Examiner tip ✓

The Celsius and Kelvin temperature scales use the same size units, so ΔT in °C is the same as ΔT in K.

Worked Example

Using an insulated coffee-cup calorimeter a student mixed 50 cm^3 each of 2.0 mol dm^{-3} HCl and 2.0 mol dm^{-3} NaOH. They both started at 23.2°C and the final temperature was 36.5°C. Calculate the enthalpy change of neutralization per mole of acid/alkali.

The temperature rise was 13.3°C, or 13.3 K.

The total mass of the solution can be taken as 100 g, so the energy required for this temperature rise is given by:

$$E = mc\Delta T$$
$$= 100 \, g \times 4.2 \, J \, g^{-1} K^{-1} \times 13.3 \, K$$
$$= 5.6 \, kJ$$

The amount of acid in 50 cm^3 of 2.0 mol dm^{-3} HCl

$$= volume \times concentration$$

$$= \frac{50 \, cm^3}{1000 \, cm^3} \times 2.0 \, mol \, dm^{-3}$$

$$= 0.10 \, mol$$

The amount of alkali in 50 cm^3 of 2.0 mol dm^{-3} NaOH is the same, i.e. 0.10 mol.

The equation for the reaction is HCl(aq) + NaOH(aq) → NaCl(aq) + H$_2$O(l)

Therefore, 0.10 mol HCl reacting with 0.10 mol NaOH will produce 0.10 mol H$_2$O.

So the enthalpy change per mole $= \dfrac{5.6 \, kJ}{0.1 \, mol}$

$$= 56.0 \, kJ \, mol^{-1}$$

This is an exothermic reaction so the enthalpy change is −56.0 kJ mol^{-1}.

Thinking Task

Will the enthalpy change of neutralization of a weak acid be different from that of a strong one?

Quick Questions

1 What assumptions are made in a coffee-cup calorimeter experiment?
2 How much energy is produced when 25 cm^3 of 1.0 mol dm^{-3} sulfuric acid is neutralized by 75 cm^3 of 1.0 mol dm^{-3} NaOH given a temperature rise of 12.1°C?
3 A butane (C$_4$H$_{10}$) gas burner was used to heat 1000 g of water in a kettle from 20°C to 100°C. If 13.5 g butane was used what is the standard enthalpy change of combustion of butane?

Using Hess's law

Hess's law states that:
- the total enthalpy change for a reaction is independent of the route taken, provided the initial and final conditions are the same.

In other words, for any chemical change, the enthalpy change is the same whatever reaction route is taken.

Hess's law is used to calculate the standard enthalpy change for reactions that are difficult to measure or that cannot be measured directly. The requirements are that.
- there are two routes between the reactants and products to be able to draw a **Hess cycle**
- the data for the enthalpy changes for one route is given.

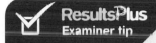

Applying Hess's law: $\Delta H_1 = \Delta H_2 + \Delta H_3$

Calculating standard enthalpy changes of formation from standard enthalpy changes of combustion

The standard enthalpy change of formation ($\Delta H_f^{\ominus}$) of a hydrocarbon fuel can be found using enthalpies of combustion ($\Delta H_c^{\ominus}$). You draw a Hess cycle linking the formation of the fuel with the combustion of the fuel, and the combustion of its constituent elements.

Worked Example

The direct route for the formation of methane is:
$$C(s) + 2H_2(g) \rightarrow CH_4(g)$$
The standard enthalpy changes of combustion, ΔH_c°, for carbon, hydrogen and methane can all be determined experimentally.

Note that under standard conditions the normal state of water is liquid – the gas condenses.

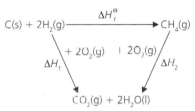

Hess cycle for the formation of methane

From the Hess cycle, we have:
$$\Delta H_f^{\circ} = \Delta H_1 - \Delta H_2$$
$$\Delta H_1 = \Delta H_c^{\ominus}[C(graphite)] + 2(\Delta H_c^{\ominus}[H_2(g)])$$
$$\Delta H_2 = \Delta H_c^{\ominus}[CH_4(g)]$$

Using the following data:
$$\Delta H_c^{\ominus}[C(graphite)] = -393.5 \text{ kJ mol}^{-1}$$
$$\Delta H_c^{\ominus}[H_2(g)] = -285.8 \text{ kJ mol}^{-1}$$
$$\Delta H_c^{\ominus}[CH_4(g)] = -890.3 \text{ kJ mol}^{-1}$$

We have:
$$\Delta H_f^{\ominus}[CH_4(g)] = (-393.5 \text{ kJ mol}^{-1}) + 2(-285.8 \text{ kJ mol}^{-1})$$
$$- (-890.3 \text{ kJ mol}^{-1})$$
$$= -74.8 \text{ kJ mol}^{-1}$$

In any Hess cycle the equations must be balanced. In this case we added $2O_2$ to both sides.

You also have to make sure that you include *all* the enthalpies – in this case *two* times $\Delta H_c^{\ominus}[H_2(g)]$ because the balanced equation for the combustion of hydrogen is $2H_2(g) + O_2(g) \rightarrow 2H_2O(g)$.

Calculating standard enthalpy changes of reaction

$CuSO_4$ is the (nearly) white anhydrous form of copper(II) sulfate. It turns blue on the addition of water – but it would be very difficult to add *exactly* the right amount of water to just hydrate all the powder. However, you can measure the enthalpy changes when you dissolve the anhydrous and hydrated forms of copper(II) sulfate separately in water, and use a Hess cycle to find $\Delta H_f^{\ominus}$ for the formation of $CuSO_4.5H_2O$.

From the Hess cycle we have $\Delta H_3 = \Delta H_1 - \Delta H_2$

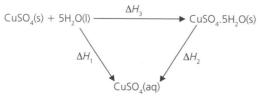

Hess cycle for the hydration of copper(II) sulfate

It makes the calculations simpler if you dissolve the same molar quantities to make the concentration $1\,mol\,dm^{-3}$, for example $0.05\,mol\,CuSO_4$ dissolved in $50\,cm^3$ water and $0.05\,mol\,CuSO_4.5H_2O$ dissolved in $45.5\,cm^3$ water (the missing $4.5\,cm^3$ come from the water of crystallization in the hydrated form).

The enthalpy changes for the two reactions can be measured using a coffee-cup calorimeter:

$$CuSO_4(s) + aq \rightarrow CuSO_4(aq)\ \Delta H_1$$
$$CuSO_4.5H_2O(s) + aq \rightarrow CuSO_4(aq)\ \Delta H_2$$

If this type of experiment is repeated a number of times, the conditions and systematic errors will be the same. Thus **reliability** is improved, **accuracy** is not.

Worked Example

To determine the enthalpy change of above we have:
$$\Delta H_3 = \Delta H_1 - \Delta H_2$$
given that $\Delta H_1 = -66.1\,kJ\,mol^{-1}$ and that $\Delta H_2 = +11.3\,kJ\,mol^{-1}$
$$\Delta H_3 = (-66.1\,kJ\,mol^{-1}) - (+11.3\,kJ\,mol^{-1})$$
$$= -77.4\,kJ\,mol^{-1}$$

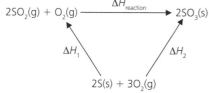

Hess cycle for the reaction of sulfur dioxide with oxygen

ResultsPlus
Watch out!

$\Delta H_f^{\ominus}$ of an element in its standard state is zero. For oxygen $\Delta H_f^{\ominus}$ is for the reaction:
$$O_2(g) \rightarrow O_2(g)$$
i.e. no reaction, therefore no enthalpy change.

Worked Example

Sulfur dioxide reacts with oxygen, in the presence of a catalyst, to form sulfur trioxide:
$$2SO_2(g) + O_2(g) \rightarrow 2SO_3(s)$$
$\Delta H_f^{\ominus}[SO_2(g)] = -297\,kJ\,mol^{-1}$
$\Delta H_f^{\ominus}[SO_3(s)] = -441\,kJ\,mol^{-1}$

Use the enthalpy changes of formation to calculate the enthalpy change of reaction.

From the Hess cycle we have $\Delta H_{reaction} = -\Delta H_1 + \Delta H_2$
$$= \Delta H_2 - \Delta H_1$$
Now $\Delta H_1 = 2\Delta H_f^{\ominus}[SO_2(g)]$ and $\Delta H_2 = 2\Delta H_f^{\ominus}[SO_3(s)]$
So, $\Delta H_{reaction} = 2(-441\,kJ\,mol^{-1}) - 2(-297\,kJ\,mol^{-1})$
$$= -288\,kJ\,mol^{-1}$$

? Quick Questions

1 **a** How much energy is transferred when $3.0\,g\,MgSO_4(s)$ is dissolved in $45\,g$ water to give a temperature rise of $11.3°C$?
   b How much energy is transferred per mole of $MgSO_4(s)$?
2 **a** How much energy is transferred when $6.2\,g\,MgSO_4.7H_2O(s)$ is dissolved in $42\,g$ water to give a temperature drop of $1.4°C$?
   b How much energy is transferred per mole $MgSO_4.7H_2O(s)$?
3 Use Hess's law to calculate the standard enthalpy change of hydration of $MgSO_4(s)$.

Bond enthalpy

Bond dissociation enthalpy is the energy associated with 1 mole of a particular bond. It is often referred to as **bond enthalpy**. It is the energy:
- required to break 1 mole of bonds
- released on making 1 mole of bonds.

For example, for the reaction $Cl_2(g) \rightarrow 2Cl(g)$ $\Delta H = +243.4 \, kJ \, mol^{-1}$. Note that this is positive – energy is needed to break Cl—Cl bonds.

Mean bond enthalpy is the mean bond dissociation enthalpy for a particular bond in a range of compounds – it can be used to calculate the enthalpy change for reactions.

Using mean bond enthalpies

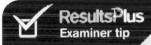

ResultsPlus
Examiner tip

Always write a chemical equation for this purpose using displayed formulae – you will be able to see all the bonds involved and not make careless mistakes.

Worked Example

Ethanol is used as a fuel. The enthalpy change of combustion can be determined experimentally and also theoretically from mean bond enthalpy data. The reaction is:

$$C_2H_5OH(l) + 3O_2(g) \rightarrow 2CO_2(g) + 3H_2O(g)$$

Use these bond enthalpies to calculate the enthalpy change of combustion.

Bond	$\Delta H/kJ \, mol^{-1}$	Bond	$\Delta H/kJ \, mol^{-1}$
C—C	347	O—H	464
C—H	413	O=O	498
C—O	358	C=O	805

The enthalpy change for this reaction, in this case $\Delta H_c^{\ominus}$, can be calculated from:

(energy needed to break bonds) − (energy released when bonds form)

The displayed formulae look like this:

- Bonds broken: (left-hand side of equation)
  $1 \times C$—$C = 1 \times 347$ $1 \times O$—$H = 1 \times 464$
  $5 \times C$—$H = 5 \times 413$ $3 \times O$=$O - 3 \times 498$
  $1 \times C$—$O = 1 \times 358$
  Total energy to break bonds = $4728 \, kJ \, mol^{-1}$

- Bonds made: (right-hand side of equation)
  $4 \times C$=$O = 4 \times 805$
  $6 \times O$ $H = 6 \times 464$
  Total energy released = $6004 \, kJ \, mol^{-1}$

Therefore, enthalpy change of reaction = $4728 \, kJ \, mol^{-1} - 6004 \, kJ \, mol^{-1}$
= $-1276 \, kJ \, mol^{-1}$

ResultsPlus
Watch out!

Breaking bonds is endothermic (ΔH is positive) and making bonds is exothermic (ΔH is negative). Be careful to check and include the sign in your final answer.

Enthalpy changes calculated from average bond enthalpy values are usually slightly different from those determined experimentally. Compare the accepted value for $\Delta H_c^{\ominus}[C_2H_5OH(l)]$ of $-1367 \, kJ \, mol^{-1}$ with that found in the example above. The difference between calculated and experimental values arises because:
- the calculations use *average* (mean) bond enthalpies
- bond enthalpies only apply to substances in the gaseous state
- as this is a liquid, the energy of vapourization must be added.

Bond enthalpies in Hess cycle calculations

Mean bond enthalpy values can be applied to Hess's law and used to calculate the enthalpy change of reactions.

The enthalpy change of formation of ethanol vapour from its elements can be estimated by applying the standard enthalpies of atomization and bond enthalpies to a Hess cycle. Given the data:

$$\Delta H_{at}^{\ominus} \, [\text{C(graphite)}] = +718 \, \text{kJ mol}^{-1}$$
$$\Delta H_{at}^{\ominus} \, [\text{H}_2\text{(g)}] = +218 \, \text{kJ mol}^{-1}$$
$$\Delta H_{at}^{\ominus} \, [\text{O}_2\text{(g)}] = +249 \, \text{kJ mol}^{-1}$$

and the mean bond enthalpies given above.

Firstly, draw the Hess cycle:

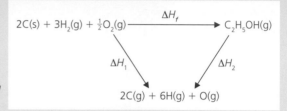

Hess cycle for the formation of ethanol vapour

Applying Hess's law, $\Delta H_f[\text{C}_2\text{H}_5\text{OH(g)}] = \Delta H_1 + \Delta H_2$

where

ΔH_1 = sum of enthalpies of atomization of 2C(s), $3\text{H}_2\text{(g)}$ and $\frac{1}{2}\text{O}_2\text{(g)}$.
ΔH_2 = sum of bond enthalpies (negative values – bonds are made)
$\Delta H_1 = [2 \times (+718 \, \text{kJ mol}^{-1})] + [3 \times (+218 \, \text{kJ mol}^{-1})]$
$\qquad\quad + \left[\frac{1}{2} \times (+249 \, \text{kJ mol}^{-1})\right]$
$\qquad = +2215 \, \text{kJ mol}^{-1}$
$\Delta H_2 = (1 \times -347 \, \text{kJ mol}^{-1}) + (5 \times -413 \, \text{kJ mol}^{-1}) + (1 \times -358 \, \text{kJ mol}^{-1})$
$\qquad\quad + (1 \times -464 \, \text{kJ mol}^{-1})$
$\qquad = -3234 \, \text{kJ mol}^{-1}$

Hence $\Delta H_f[\text{C}_2\text{H}_5\text{OH(g)}] = +2215 \, \text{kJ mol}^{-1} + (-3234 \, \text{kJ mol}^{-1})$
$\qquad\qquad\qquad\qquad\qquad = -1019 \, \text{kJ mol}^{-1}$

Watch out!

This is *not* the standard enthalpy change of formation of ethanol. Standard enthalpies apply only to substances in their *standard states*. Ethanol is a liquid at 1 atm and 298 K. This calculation is for the formation of the *vapour* – bond enthalpies only apply to substances in the gaseous state.

Bond enthalpy and stability

The larger the bond enthalpy, the stronger the bond – for example, ΔH *for the* $\text{N} \equiv \text{N}$ bond is $945 \, \text{kJ mol}^{-1}$. It is very strong – nitrogen is relatively unreactive.

ΔH for the $\text{C} - \text{H}$ bond is $435 \, \text{kJ mol}^{-1}$. This is much weaker – methane burns in air. Bond enthalpy data gives an indication of:
- the ease with which a bond will break, and which bond will break first in a reaction
- how fast a reaction might be.

Thinking Task

Consider the reaction
$\text{C}_{\text{(diamond)}} \rightarrow \text{C}_{\text{(graphite)}} \; \Delta H^{\ominus}$
$= -1.9 \, \text{kJ mol}^{-1}$.
What does this tell us about the reaction? What does it not tell us about the reaction? Are diamonds slowly turning into graphite?

Quick Questions

1 a Use the bond enthalpy data in this section to estimate the enthalpy change for the reaction of propene with steam to produce propan-2-ol:

$$\text{C}_3\text{H}_6\text{(g)} + \text{H}_2\text{O(g)} \rightarrow \text{C}_3\text{H}_7\text{OH(l)}$$

You also need to use $\Delta H \, [\text{C} = \text{C}] = +612 \, \text{kJ mol}^{-1}$.
   b Why is this value not the standard enthalpy change for the reaction?
2 Use an appropriate data source to find the bond dissociation enthalpies for $\text{F}_2\text{(g)}$, $\text{Cl}_2\text{(g)}$, $\text{Br}_2\text{(g)}$ and $\text{I}_2\text{(g)}$. Does consideration of these enthalpies alone explain the trends in reactivity of the halogens?

Topic 2: Energetics and enthalpy changes checklist

By the end of this topic you should be able to:

Revision spread	Checkpoints	Specification section	Revised	Practice exam questions
Enthalpy changes and enthalpy level diagrams	Understand the term enthalpy change, ΔH	1.4a	☐	☐
	Draw enthalpy level diagrams showing the enthalpy change	1.4b	☐	☐
	Recall the sign of ΔH for exothermic and endothermic reactions	1.4c	☐	☐
Measuring enthalpy changes	Recall the definition of standard enthalpy changes of reaction, formation, combustion, neutralization and atomization; calculate the energy transferred in a reaction and hence the enthalpy change of the reaction	1.4d	☐	☐
	Evaluate results obtained from experiments using the expression: energy transferred in joules = mass × specific heat capacity × temperature change and comment on sources of error and assumptions made in the experiments	1.4f i and ii	☐	☐
Using Hess's law	Recall Hess's law and calculate enthalpy changes of reaction from data provided, and understand why data should be given for standard conditions	1.4e	☐	☐
	Plan and carry out an experiment where the enthalpy change cannot be measured directly, and use Hess's law to calculate enthalpy changes from data for these reactions	1.4f iii	☐	☐
Bond enthalpy	Understand the terms bond enthalpy and mean bond enthalpy, and use bond enthalpies in Hess cycle calculations and recognise their limitations Understand that bond enthalpy data gives some indication about which bond will break first in a reaction, how easy or difficult it is and, therefore, how rapidly a reaction will take place at room temperature	1.4g	☐	☐

ResultsPlus
Build Better Answers

1 a Define the term standard enthalpy change of combustion. (3)

✓ Examiner tip

The enthalpy change when 1 mole of a substance (1) burns completely in air/oxygen (1) in conditions of 1 atm pressure and specified temperature (298 K) (1)

Many students fail to gain full marks when defining standard enthalpy changes. You must include the proper definition of the change, 1 mole of the correct material and standard conditions. You should also give values, not just 'room temperature and pressure'.

b The following standard enthalpy changes of combustion are needed to calculate the standard enthalpy change of formation of ethanol, C_2H_5OH.

Substance	Standard enthalpy change of combustion /$kJ\,mol^{-1}$
carbon, C(s, graphite)	−394
hydrogen, H_2(g)	−286
ethanol, C_2H_5OH(l)	−1371

i Complete the Hess's law cycle by filling in the boxes and labelling the arrows with the enthalpy changes. (3)

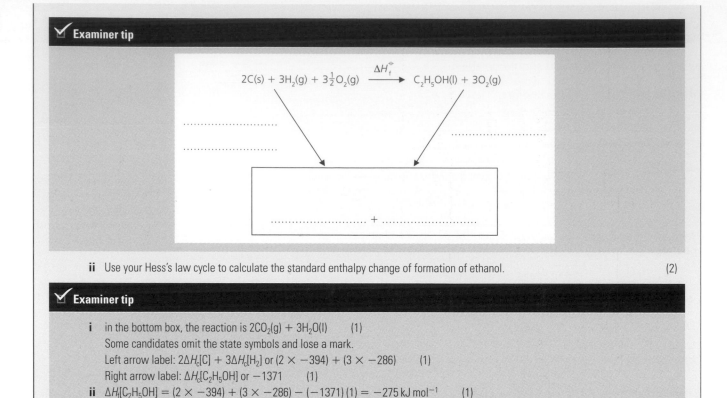

ii Use your Hess's law cycle to calculate the standard enthalpy change of formation of ethanol. (2)

i in the bottom box, the reaction is $2CO_2(g) + 3H_2O(l)$ (1)
Some candidates omit the state symbols and lose a mark.
Left arrow label: $2\Delta H_c[C] + 3\Delta H_c[H_2]$ or $(2 \times -394) + (3 \times -286)$ (1)
Right arrow label: $\Delta H_c[C_2H_5OH]$ or -1371 (1)
ii $\Delta H_f[C_2H_5OH] = (2 \times -394) + (3 \times -286) - (-1371)$ (1) $= -275\,kJ\,mol^{-1}$ (1)
Some candidates omit the unit and lose the second mark.

(From Edexcel Unit test 2 Q5, June 07)

Practice exam questions

1 The combustion of propanone is given by the equation:

$$CH_3COCH_3(g) + 4O_2(g) \rightarrow 3CO_2(g) + 3H_2O(g)$$

Using the mean bond enthalpies (in $kJ\,mol^{-1}$) in the table earlier in this section, calculate the enthalpy change for the reaction. (7)

2 A Hess cycle for the formation of propane is shown below.

$$3C(\text{graphite}) + 4H_2(g) \xrightarrow{\Delta H_f^{\ominus}} C_3H_8(g)$$

ΔH_1 $+5O_2(g)$ $+5O_2(g)$ ΔH_2

$$3CO_2(g) + 4H_2O(g)$$

Use the diagram and the following data to calculate $\Delta H_f^{\ominus}$ of propane.
Include a sign in your answer. (3)

$$\Delta H_c^{\ominus}[C(s)\text{graphite}] = -394\,kJ\,mol^{-1}$$
$$\Delta H_c^{\ominus}[H_2(g)] = -286\,kJ\,mol^{-1}$$
$$\Delta H_c^{\ominus}[C_3H_8(l)] = -2219\,kJ\,mol^{-1}$$

Mass spectrometry

Mass spectrometry of elements

Make sure you know the following – you have seen them before!

- The **mass number (A)** of an atom is given by the sum of protons (Z) and neutrons in that atom.
- The **atomic number (Z)** is given by the number of protons in an atom.
- **Isotopes** are atoms of the same element (same number of protons) with different numbers of neutrons, therefore they have different mass numbers.
- **Relative atomic mass** (RAM) is the average (weighted) mass of an element's isotopes relative to 1/12 the mass of a ^{12}C atom.
- **Relative isotopic mass** is the mass of an atom of an isotope relative to 1/12 the mass of a ^{12}C atom.

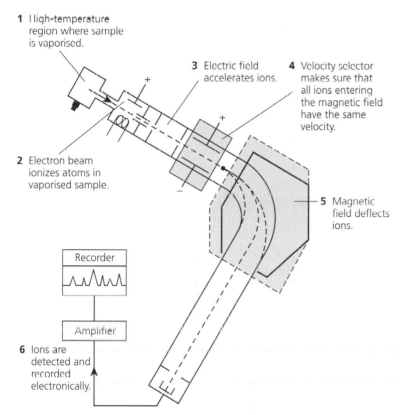

1 High-temperature region where sample is vaporised.

3 Electric field accelerates ions.

4 Velocity selector makes sure that all ions entering the magnetic field have the same velocity.

2 Electron beam ionizes atoms in vaporised sample.

5 Magnetic field deflects ions.

Recorder

Amplifier

6 Ions are detected and recorded electronically.

A mass spectrometer

The **mass spectrometer** measures the masses of positive ions formed from atoms:

1 The sample is vaporised.
2 The vapour is ionized by bombardment with high-energy electrons and electrons are knocked out of the atoms.
3 The (positive) ions are accelerated by an electric field.
4 The ions have their velocities equalised in a velocity selector.
5 The ions are deflected in a magnetic field.
6 The field is steadily increased – only ions of a particular $\dfrac{\text{mass}}{\text{charge}}$ ratio (m/z) pass through at any one time and reach the detector.

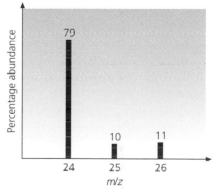

The mass spectrum of magnesium

Most of the ions formed have a charge of +1 (loss of one electron), so m/z corresponds to the mass of the ion.

A **mass spectrum** shows us the masses of the ions detected and their relative abundance – i.e. the **isotopic composition** of an element.

From this, the relative atomic mass of the element can be calculated (see page 8).

Mass spectrometry and compounds

The mass spectrometer can also be used to measure the relative molecular mass of a compound. Remember, the **relative formula mass** is the sum of the relative atomic masses of all the atoms or ions within the formula. For a covalent molecule this is the **relative molecular mass**.

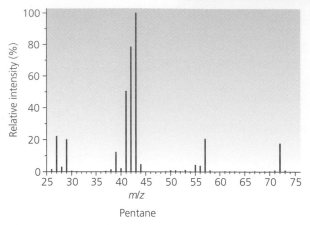

Pentane

The mass spectrum of pentane

When the vaporised sample is bombarded with high-energy electrons, not only are electrons knocked off, but the molecule is **fragmented**. The mass spectrum of an organic molecule comprises a number of **peaks** corresponding to the various fragments of differing masses produced during electron bombardment. The peak with the largest *m/z* value is due to the **molecular ion**, M$^+$ (also called parent ion). It is formed when a sample molecule loses just one electron and is not fragmented.

The mass spectrum of pentane has a peak at *m/z* = 72 for the molecular ion $C_5H_{12}^+$. Any compound with the same largest *m/z* value will have a relative molecular mass of 72.

The uses of mass spectrometry

Mass spectrometry has a wide range of applications, some of which are:
- radioactive dating of archaeological specimens – the percentage abundance of radioactive isotopes is measured
- space research – identification of molecules detected by planetary probes
- detection of illegal drugs in athletes' urine samples, such as anabolic steroids – each molecule can be identified by its unique mass spectrum
- identifying molecules with potential for use as drugs in the pharmaceutical industry – the likely structure of the new compounds can be identified from its mass spectrum

Thinking Task

What assumptions do you think are made in radioactive dating?

? Quick Questions

1 Why can mass spectrometry be used to detect anabolic steroids in sport?
2 **a** What are the relative formula masses of the following compounds?

    CO_2, NaOH, $CaCO_3$, CH_3OCH_3, Na_2SO_4, CH_3OH.

   b Which are relative molecular masses?
3 A sample of magnesium contains three isotopes of mass numbers 24, 25 and 26. In terms of sub-atomic particles, state **one** similarity and **one** difference between these isotopes.
4 The mass spectrum of a sulfur sample showed three peaks. The first peak was at *m/z* = 32 and had an abundance of 94.93%; the second peak was at *m/z* = 33 and had an abundance of 0.76%; the third peak was at *m/z* = 34 and had an abundance of 4.29%. Calculate the relative atomic mass of the sulfur sample to three significant figures.
5 The mass spectrum of propane shows a large peak at 29 and smaller peaks at 27, 28 and 44. What is the relative molecular mass of propane?

Ionization energy and electron shells

Ionization energy indicates the amount of energy required to remove an electron from an atom. This is an endothermic process. It is measured as the energy required to remove 1 mole of electrons from 1 mole of atoms in the gaseous state.

- The **first ionization energy** (1st I.E.) gives an idea of how easily an atom loses an electron to form a 1+ ion. For example, the first ionization energy of sodium is represented as the equation:

$$Na(g) \rightarrow Na^+(g) + e^- \qquad \text{1st I.E.} = +496 \, kJ \, mol^{-1}$$

- The **second ionization energy** (2nd I.E.) is the energy required to remove an electron from a 1+ ion to form a 2+ ion. For example:

$$Na^+(g) \rightarrow Na^{2+}(g) + e^- \qquad \text{2nd I.E.} = +4563 \, kJ \, mol^{-1}$$

The total energy required to form a 2+ ion from an atom is the sum of the first and second ionization energies.

Evidence for electron shells

The series of successive ionization energies for an element (see diagram) provide evidence for the existence of **electron (quantum) shells** (or energy levels) around the nucleus.

Logarithms are used to plot ionization energies because they range over several orders of magnitude.

- The first ionization energy is the lowest, because an electron in an **outer shell**, furthest from the nucleus, is removed first.
- The very large increase between first and second ionization energies of sodium arises because the second electron has to be removed from a shell that is closer to the nucleus.
- Electrons closer to the nucleus have fewer electron shells between them and the nucleus – they are less 'shielded' from the positive attraction of the protons.
- Electron shells closer to the nucleus are attracted towards it more strongly.
- There is a big jump between the 9th and 10th ionization energies for sodium. The 10th and 11th electrons to be removed from sodium are in the shell closest to the nucleus and are very strongly attracted to it.
- You can work out which group an element is in from where the first big jump is in its successive ionization energies.

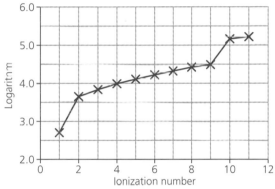

Successive ionization energies of sodium using a logarithm scale

Evidence for electron sub-shells

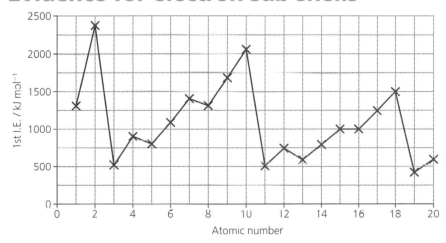

First ionization energies for the first 20 elements

The first ionization energy of successive elements provides evidence for electron **sub-shells**. For example, removing the outer electron from nitrogen ($Z = 7$) involves breaking into a half-full p sub-shell. This requires more energy than removing the outer electron from oxygen ($Z = 8$), which leaves a half-full sub-shell.

Generally, ionization energies increase across a period, such as from lithium ($Z = 3$) to neon ($Z = 10$) due to greater force of attraction between the nucleus and outer electrons. However, ionization energy decreases slightly between beryllium ($Z = 4$) and boron ($Z = 5$) and between nitrogen ($Z = 7$) and oxygen ($Z = 8$). This suggests that a given shell is divided into sub-shells that can have slightly different energies.

The reasons for the different energy levels of sub-shells within a shell can be explained by the rules for shell filling and electron pairing.

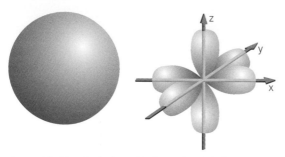

An s orbital is spherical; p orbitals are dumb-bell shaped

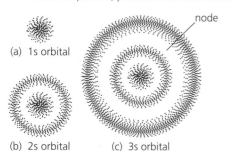

(a) 1s orbital

(b) 2s orbital   (c) 3s orbital

Electron density maps for s orbitals

Orbitals and sub-shells

An electron **orbital** is a region of space, centred on the nucleus, in which it is most likely to be found.

Each sub-shell is made up of one type of orbital – **s**, **p**, **d** or **f**. An s sub-shell (in all shells) has one s orbital, the p sub-shell (in the 2nd and later shells) has three p orbitals and the d sub-shell (in the 3rd and later shells) has five d orbitals. The orbitals have distinctive shapes – s orbitals are spherical, p orbitals are dumb-bell shaped.

Each orbital can contain a maximum of two electrons (of opposite **spin**) – so each s sub-shell can hold 2 electrons, each p sub-shell 6 electrons and each d sub-shell 10 electrons.

Electron density maps of electron orbitals show how the **electron cloud** is distributed within the orbital.

All the orbitals in a given sub-shell are at the same energy level, but the 2s sub-shell is at a different energy to the 2p sub-shell.

Writing electronic configurations

The **electronic configuration** of an atom specifies the number of electrons in each electron shell or sub-shell. Electronic configurations for atoms in their **ground state** (lowest energy state) generally reflect the order in which shells and sub-shells are filled. The lowest energy shells are filled first –1s, 2s, 2p, 3s, 3p, 4s, 3d, 4p.

Electrons populate orbitals singly before pairing up (Hund's rule). This explains why removing the outer electron from boron requires less energy than removing the outer electron from beryllium – the outermost electron goes to a new p sub-shell rather than filling the s sub-shell. The p sub-shell is at a higher energy level than the s sub-shell, so less energy is required to remove an electron from it.

The electronic configuration for krypton ($Z = 36$) is: $1s^2 2s^2 2p^6 3s^2 3p^6 3d^{10} 4s^2 4p^6$, where the superscript shows the number of electrons in that sub-shell. This can also be shown in a diagram like this:

Electron box diagram for krypton – arrows represent electrons of opposite spin

? Quick Questions

1   Use your data booklet to find the energy required to form 1 mole of Mg^{2+} ions in the gaseous state.
2   What is the electronic configuration of arsenic?
3   What is special about full or half-full sub-shells?
4   An element has successive ionization energies 906, 1763, 14 855 and 21 013 kJ mol^{-1}. What evidence is there that this element is in Group 2?

Electronic configurations and periodic properties

Electronic configurations

The electronic structure of an atom determines the chemical properties of that element. The outer shell electrons are those which take part in chemical reactions (bonding) and have most influence over the chemistry of the atom.

The Periodic Table arranges the elements in order of their atomic number (Z). The **group number** in the Periodic Table is the same as the number of electrons in the outer shell of the elements in that group:
- Group 1 elements (Li, Na, K etc.) have 1 outer electron
- Group 7 elements (F, Cl, Br etc.) have 7 outer electrons.

The Periodic Table is also divided into blocks – **s block**, **p block** and **d block** – in which the last electrons to be added occupy specific orbitals – s, p or d, respectively. This should help you to predict electronic configurations from the order of electron filling of sub-shells. For example, the outer electron in arsenic (p block) occupies a p orbital.

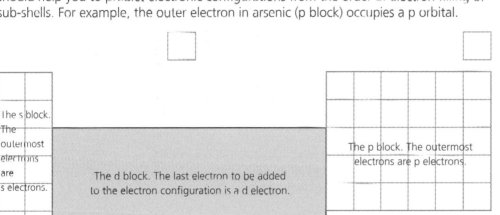

The s block. The outermost electrons are s electrons.

The d block. The last electron to be added to the electron configuration is a d electron.

The p block. The outermost electrons are p electrons.

s block   d block   p block

The s, p and d blocks of the Periodic Table

- The s block elements have 1 or 2 outer electrons and are reactive metals.
- The d-block elements are **transition metals** – they have similar chemical properties because the electrons are not being added to outer electron shells; the d orbitals are *inside* the s orbitals.

It is the outer electrons that have most influence on chemical properties.

Periodic properties

A **periodic** property has a trend which repeats across periods. For example, the first ionization energy tends to increase across a period and decreases down a group.

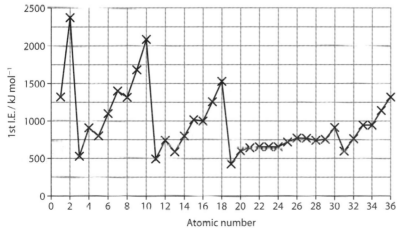

The first ionization energy is a periodic property

Going right across a period, the outer electron is drawn closer to the increasingly positive nucleus. There is also constant shielding from the nucleus by the full inner shell(s). Overall there is a greater force of attraction between the nucleus and outer electron, and it becomes more difficult to remove it from the atom.

Going down a group, the outer electron is increasingly further from the nucleus and electron shielding by the additional inner shells reduces the attraction to the nucleus – it becomes easier to remove it from the atom.

Melting temperature is another **periodic** property of elements.

The melting temperature of an element depends on its structure and bonding (see Topic 4):

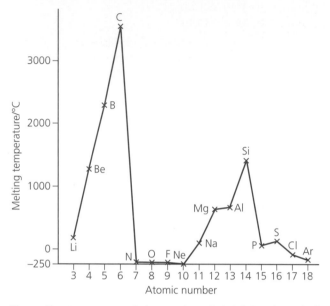

The melting temperature of elements in periods 2 & 3 varies periodically

- metallic structures (e.g. metals in Groups 1, 2 and 3) are strong – it takes a lot of energy to break the bonds that hold the atoms/ions together
- giant molecular structures (e.g. C and Si in Group 4) are held together by covalent bonds which are strong – it is very difficult to melt these elements
- simple molecular structures (e.g. S_8 and Cl_2) are held together by weak forces of attraction between the molecules – it requires little energy to overcome these intermolecular forces of attraction, the melting temperatures are low.

Melting temperatures rise from Group 1 to Group 2 to Group 3 because the number of shared delocalized electrons increases, which increases the strength of metallic bonding.

Quick Questions

1   Why do the outer electrons of an atom have the most influence over its chemistry?
2   Why is the first ionization energy of sodium larger than that of potassium, but smaller than that of argon?
3   Why is the melting temperature of phosphorus much lower than that of diamond?

Topic 3: Atomic structure and the Periodic Table checklist

By the end of this topic you should be able to:

Revision spread	Checkpoints	Specification section	Revised	Practice exam questions
Mass spectrometry	Recall the definitions of relative atomic mass, relative isotopic mass and relative molecular mass and understand that they are measured relative to 1/12 the mass of a ^{12}C atom	1.5a	☐	☐
	Understand the basic principles of a mass spectrometer and interpret data from a mass spectrometer to: (i) deduce the isotopic composition of a sample of an element, e.g. polonium (ii) deduce the relative atomic mass of an element (iii) deduce the relative molecular mass of a compound	1.5b	☐	☐
	Describe some uses of mass spectrometers	1.5c	☐	☐
Ionization energy and electron shells	Recall and understand the definition of ionization energies of gaseous atoms and that they are endothermic processes	1.5d	☐	☐
	Recall that ideas about electronic structure developed from: (i) evidence provided by successive ionization energies (ii) evidence provided by the first ionization energy of successive elements	1.5e	☐	☐
	Describe the shapes of electron density plots (or maps) for s and p orbitals	1.5f	☐	☐
	Predict the electronic structure and configuration of atoms of the elements from hydrogen to krypton inclusive using the 1s... notation and electron-in-boxes notation	1.5g	☐	☐
Electron configurations and periodic properties	Understand that electronic structure determines the chemical properties of an element	1.5h	☐	☐
	Recall that the Periodic Table is divided into blocks, such as s, p and d	1.5i	☐	☐
	Represent data in a graphical form for elements 1 to 36 and use this to explain periodic properties	1.5j	☐	☐
	Explain trends in the following properties of the elements from Periods 2 and 3: (i) melting temperature of the elements based on structure and bonding (ii) ionization energy	1.5k	☐	☐

ResultsPlus
Build Better Answers

1 a Define relative isotopic mass. (2)

✓ Examiner tip

Most students can state that the mass is relative to 1/12 the mass of ^{12}C (1), however the second mark is for saying that this is for the mass of an *atom* (of the isotope) (1). Be careful to answer the question as it is written – don't miss out on marks for writing a definition of an isotope, or of *relative* atomic mass.

b The following data were obtained from the mass spectrum of a sample of chromium.

Calculate the relative atomic mass of this sample of chromium. Give your answer to four significant figures. (2)

Relative isotopic mass	Percentage abundance
49.95	4.345
51.94	83.79
52.94	9.501
53.94	2.364

There is one mark for the correct answer, but also one mark for giving the answer to four significant figures.

$$RAM = \frac{(49.95 \times 4.345) + (51.94 \times 83.79) + (52.94 \times 9.501) + (53.94 \times 2.364)}{100} \quad (1)$$

$$= 51.9958$$

$$= 52.00 \quad (1)$$

Many students round up 51.9958 incorrectly to 51.99, or even 52.10. Make sure you understand that zeros after a decimal place are significant figures.

c   Copy and complete the electronic configuration of an iron atom, atomic number 26.   (2)

1s	2s	2p			3s	3p			3d					4s
↑↓	↑↓	↑↓	↑↓	↑↓										

Basic answers get the number of electrons correct (26) and also show two electrons in the 4s sub-shell (remember that the 4s orbital fills before the 3d). For both marks, you need to show that the electrons fill the 3d orbitals singly before pairing in any one orbital.

1s	2s	2p			3s	3p			3d					4s
					↑↓	↑↓	↑↓	↑↓	↑↓	↑	↑	↑	↑	↑↓

(Adapted from Edexcel Unit test 1 Q1, June 08)

Practice exam questions

1   Put the species S^+, S and S^- in order of increasing first ionization energy, starting with the lowest.   (1)

2   The first five ionization energies of an element, Z, are 790, 1600, 3200, 4400, 16 100 kJ mol^{-1}

    In which group of the Periodic Table is Z found?   (1)

    A 2   **B** 3   **C** 4   **D** 5

3   **a**   Complete the electronic configuration for calcium, Ca.

          $1s^2 \ldots$ (1)

    b   **i**   Define the term first ionization energy.   (3)

         ii   Explain why the first ionization energy of calcium is lower than that of magnesium.   (3)

4   **a**   Describe briefly how positive ions are formed from gaseous atoms in a mass spectrometer.   (2)

    b   What is used to accelerate the ions in a mass spectrometer?   (1)

    c   What is used to deflect the ions in a mass spectrometer?   (1)

                           (From Edexcel Unit 1 test Q20, Jan 09)

5   Which of the following are reasonable values for the first ionization energies of elements from nitrogen to sodium? (in kJ mol^{-1})   (1)

    A 520, 900, 801, 1086, 1402

    B 1402, 1086, 801, 900, 520

    C 1402, 1314, 1681, 2081, 496

    D 1402, 1086, 801, 900, 1314

Iconic bonding

The evidence that ions exist

- **Physical properties** of ionic compounds: high melting temperatures, showing strong forces of attraction between ions, soluble in polar solvents, conduct electricity when molten or in aqueous solution.
- **Electron density maps** of compounds produced from X-ray diffraction patterns show zero electron density between ions – meaning complete electron transfer.
- Migration of ions in **electrolysis**. For example, electrolysis of green aqueous copper(II) chromate(VI) attracts a yellow colour (chromate(VI) ions) to the anode and a blue colour (copper(II) ions) to the cathode.

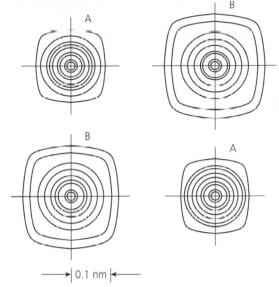

→| 0.1 nm |←

Electron density map for sodium chloride – clearly showing separate ions

Formation of ions

An ion is formed when an atom gains or loses one or more electrons. For example, a copper atom loses two electrons:

$$Cu(g) \rightarrow Cu^{2+}(g) + 2e^-$$

to form a copper **cation**. A chlorine atom gains one electron:

$$Cl(g) + e^- \rightarrow Cl^-(g)$$

to form a chloride **anion**.
- A positive ion is called a **cation**, it is attracted towards the cathode in electrolysis.
- A negative ion is called an **anion**, it is attracted towards the anode in electrolysis.

When ions are formed they tend to have a full outer shell, i.e. eight electrons. This is called the **octet rule**. Ions with full outer shells have the same electronic configurations as noble gases (Group 0) – for example, a Ca^{2+} ion has the same electronic configuration as argon; they are **isoelectric**.

ResultsPlus
Watch out!

The formation of an ion is *not* the same as the formation of an ionic bond. Ions *are* formed as ionic bonds are made, but they can also be formed by other means, e.g. electron bombardment in a mass spectrometer.

Ionic dot-and-cross diagrams

Only electrons in the outer shell of the atom or ion are shown in these. Electrons are drawn:
- at the four points of the compass – north, south, east and west
- paired up until there remains just an odd one.

The reactions of elements to form ionic compounds can be represented by **dot-and-cross diagrams**. The electrons from one reactant are usually shown with crosses and the electrons from the other reactant with dots.

The formation of sodium chloride and calcium fluoride by electron transfer

Electrons are *transferred* from one atom to another to form ions, each with an outer shell of eight electrons. You must show the charge on each ion in the compound. The new outer shell of the Na⁺ and Ca²⁺ cations was an inner shell in the atoms – it is not usually shown in dot-and-cross diagrams (although you may be asked to show *all* electrons).

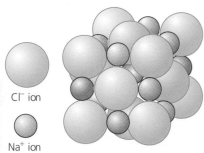

Cl⁻ ion

Na⁺ ion

A small part of the giant ionic lattice of sodium chloride

Ionic bonding and lattices

An **ionic bond** is an omnidirectional **electrostatic force** of attraction between oppositely charged ions.

- The forces of attraction are equal in all directions.
- In **ionic compounds** each ion is surrounded by ions of the opposite charge.
- Ionic compounds form **giant ionic lattices** in the solid state (also called an ionic crystal).

Trends in ionic radii

The **ionic radius** is the radius of an ion in its crystal form.

- Cations are smaller than the original atom since the atom loses electrons. Usually a whole electron shell has been lost, and the remaining electrons are also pulled in towards the nucleus more strongly.
- Anions are larger than the original atom since the atom gains electrons and there is more repulsion in the electron cloud.

Going down a group in the Periodic Table, the ions become larger – the number of shells is increasing.

Isoelectronic ions have different ionic radii:

- the additional electrons in anions make the ions larger because there is greater repulsion and all the electrons are less tightly bound than in the atom
- the loss of electrons to form cations means the nucleus attracts those electrons that remain more strongly. For example, for the electronic configuration $1s^2 2s^2 2p^6$, the order of size is
  $N^{3-} > O^{2-} > F^- > Ne > Na^+ > Mg^{2+} > Al^{3+}$.

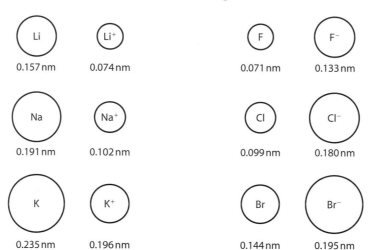

Ionic radius increases down a group – cations are smaller than their atoms

Ionic radius increases down a group – anions are larger than their atoms

Quick Questions

1 In how many directions does an ionic bond act?
2 How is an ion formed?
3 Draw a dot-and-cross diagram to represent magnesium chloride.
4 **a** Using your data book, draw scale diagrams, in decreasing order of size, of the following ions: N^{3-}, O^{2-}, F^-, Na^+, Mg^{2+} and Al^{3+}.
   b What are their electronic configurations?
   c Why do the ions have different sizes?

Lattice energies and Born–Haber cycles

The formation of an ionic crystal from its elements is exothermic (energy is released). The **lattice energy** is the energy released when 1 mole of an ionic crystal is formed from its ions in the gaseous state, under standard conditions. This process can be broken down into a number of stages, each associated with a particular energy change:

- atomization of the metal
- ionization of the gaseous metal
- atomization of the non-metal
- ionization of gaseous non-metal atoms (called the electron affinity)
- forming the crystal from the gaseous ions.

Having measured each of these, and the standard enthalpy of formation of $CaCl_2$, $\Delta H_f^{\ominus}[CaCl_2]$, the lattice energy can be calculated using an enthalpy level diagram known as a **Born–Haber cycle**. This is just an application of Hess's law.

Worked Example

Calculate the lattice energy for calcium chloride: $Ca^{2+}(g) + 2Cl^-(g) \rightarrow CaCl_2(s)$.

The enthalpy changes involved in the formation of calcium chloride from calcium and chlorine are:

1. $Ca(s) \rightarrow Ca(g)$ $\Delta H_{at}^{\ominus}[Ca(s)]$ standard enthalpy change of atomization of calcium
2. $Ca(g) \rightarrow Ca^+(g) + e^-$ $\Delta H_{i_1}^{\ominus}[Ca(g)]$ 1st ionization energy of calcium
3. $Ca^+(g) \rightarrow Ca^{2+}(g) + e^-$ $\Delta H_{i_2}^{\ominus}[Ca(g)]$ 2nd ionization energy of calcium
4. $Cl_2(g) \rightarrow 2Cl(g)$ $2 \times \Delta H_{at}^{\ominus}[\frac{1}{2}Cl_2(g)]$ standard enthalpy change of atomization of chlorine
5. $2Cl(g) + 2e^- \rightarrow 2Cl^-(g)$ $2 \times \Delta H_e^{\ominus}[Cl(g)]$ 1st electron affinity of chlorine
6. $Ca(s) + Cl_2(g) \rightarrow CaCl_2(s)$ $\Delta H_f^{\ominus}$ enthalpy change of formation of $CaCl_2$

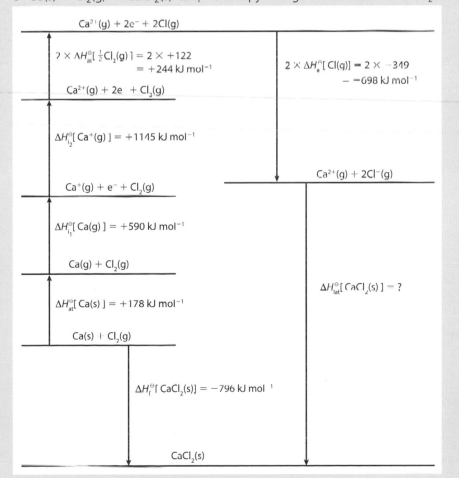

The Born–Haber cycle for the formation of calcium chloride

ResultsPlus
Watch out!

Make sure you include all the ionization energies for the metal. This means only the 1st I.E. for Group 1 metals, but the *sum* of the 1st and 2nd I.E. for Group 2 metals (not just the 2nd I.E.). Also, the standard enthalpy of atomization is associated with the production of 1 mole of gaseous atoms – when you have to produce 2 moles of atoms, you must multiply by 2.

ResultsPlus
Examiner tip

You will not have to draw a full Born–Haber cycle but you can expect to have to fill in some missing values and then calculate the remaining unknown energy (don't forget the correct sign and units). Simply move around the cycle in one direction, adding up the energies. It's exactly the same as using a Hess cycle – remember, when going in the direction of an arrow use the same sign as the energy; going against the arrow, use the opposite sign.

Using the Born–Haber cycle, the lattice energy for $CaCl_2$ is given by:

$$\Delta H_{lat}^{\ominus}[CaCl_2(s)] = [-(-698) - 244 - 1145 - 590 - 178 - 796]\,kJ\,mol^{-1}$$
$$= -2255\,kJ\,mol^{-1}$$

This is very large and indicates the strength of ionic forces of attraction between ions.

Stability of ionic compounds

Born–Haber cycles can be used to predict the relative stabilities of ionic compounds, and even if a particular formula will exist as a compound.

Worked Example

Why is calcium chloride $CaCl_2$, and not $CaCl$ or $CaCl_3$?

Enthalpy change in the process, $\Delta H^{\ominus}$/kJ mol^{-1}	Ca^+Cl^-	$Ca^{2+}(Cl^-)_2$	$Ca^{3+}(Cl^-)_3$
$Ca(s) \rightarrow Ca(g)$	+178	+178	+178
$Ca(g) \rightarrow Ca^+(g) + e^-$	+590	+590	+590
$Ca^+(g) \rightarrow Ca^{2+}(g) + e^-$		+1145	+1145
$Ca^{2+}(g) \rightarrow Ca^{3+}(g) + e^-$			+4912
$\frac{1}{2}Cl_2(g) \rightarrow Cl(g)$	+122		
$Cl_2(g) \rightarrow 2Cl(g)$		+244	
$\frac{3}{2}Cl_2(g) \rightarrow 3Cl(g)$			+366
$Cl(g) + e^- \rightarrow Cl^-(g)$	−349		
$2Cl(g) + 2e^- \rightarrow 2Cl^-(g)$		−698	
$3Cl(g) + 3e^- \rightarrow 3Cl^-(g)$			−1047
$Ca^+(g) + Cl^-(g) \rightarrow Ca^+Cl^-(s)$	−711		
$Ca^{2+}(g) + 2Cl^-(g) \rightarrow Ca^{2+}(Cl^-)_2(s)$		−2255	
$Ca^{3+}(g) + Cl^-(g) \rightarrow Ca^{3+}(Cl^-)_3(s)$			−4803

Summary of Born–Haber data for CaCl, CaCl₂ and CaCl₃

The theoretical enthalpy change of formation for each compound is found by adding the energies in each of the columns. Using the data, the enthalpy changes of formation for the three compounds are:

- $Ca(s) + \frac{1}{2}Cl_2(g) \rightarrow CaCl(s)$ $\Delta H_f^{\ominus} = -170\,kJ\,mol^{-1}$
- $Ca(s) + Cl_2(g) \rightarrow CaCl_2(s)$ $\Delta H_f^{\ominus} = -796\,kJ\,mol^{-1}$
- $Ca(s) + \frac{3}{2}Cl_2(g) \rightarrow CaCl_3(s)$ $\Delta H_f^{\ominus} = +1341\,kJ\,mol^{-1}$

The most stable compound is that with the most exothermic enthalpy of formation, $CaCl_2(s)$. The formation of $CaCl_3$ is highly endothermic, because the very high 3rd ionization enthalpy cannot be provided by the extra lattice energy.

(?) Quick Questions

1   How do you find the enthalpy change for the formation of the $Al^{3+}(g)$ ion from the element?
2   What does the general size of lattice energies tell us about ionic bonding?

Testing the ionic model

The Born–Haber cycle uses measured enthalpies to calculate lattice energies. Lattice energy can also be calculated from a model using Coulomb's law (electrostatic attraction), assuming complete electron transfer in ionic compounds, and the size of the ionic radii. Coulomb's law calculates the force of attraction between ions as a function of their charges and the distance between them.

This ionic model can be tested for different compounds by comparing experimental (Born–Haber) results with theoretical results from the model.

Compound	Lattice energy/kJ mol^{-1}	
	Born–Haber	Theoretical
NaF	918	912
NaCl	780	770
NaBr	742	735
NaI	705	687
AgF	958	920
AgCl	905	883
AgBr	891	816
AgI	889	778

The different methods of obtaining the lattice energies for the sodium halides produce very similar results. But this is not the case for the silver halides – their lattice energies are more exothermic (more negative and more stable) than theory would predict.
- Coulomb's law assumes that the ions are completely separate and spherical (not distorted).
- Experiment therefore suggests there is a degree of electron sharing, i.e. covalency, in the silver halides, while the sodium halides show (almost) pure ionic bonding.

This is supported by lower melting temperatures for silver halides than sodium halides.

Polarization of ions

Polarization of an ion is the distortion of its **electron cloud** away from completely spherical:
- a cation will distort an anion
- a cation has **polarizing power**
- an anion is **polarizable**.

ResultsPlus
Watch out!

Cations polarize. Anions are polarized.

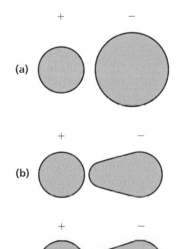

(a)

(b)

(c)

Ionic bonds can be distorted

The **polarizing power** of a cation depends on its charge density:

- a small cation is more polarizing than a larger one – the positive nucleus has more effect across the small ionic radius
- a cation with a large charge is more polarizing than one with a small charge – a large charge has more attraction than a small one.

The **polarizability** of an anion depends on its size alone:

- a large anion is easily polarized – its electron cloud is further from the nucleus and is held less tightly than on a smaller anion.

As shown above, for some ionic compounds the ionic model is good – Born–Haber lattice energies agree well with theoretical values. For some ion pairs the bonds have considerable covalent character – the agreement between experimental and theoretical lattice energies is poor.

The figure shows: **a** completely spherical and separate ions; **b** an anion distorted by a cation; and **c** distortion so great that the electron density resembles a covalent bond.

This idea is developed further in Unit 2, see page 66.

Thinking Task

When is an ionic bond not an ionic bond?

Thinking Task

At what point does an ionic bond become covalent, rather than just having some covalent character?

Quick Questions

1   What affects the polarizing power of a cation?
2   Why is a large anion polarizable?
3   How can an ionic bond have some covalent character?
4   What evidence is there to suggest that the bonding in silver iodide has some covalent character?

Covalent bonding

Formation of covalent bonds

A **covalent bond** is formed when a pair of electrons is **shared** between two atoms. This happens when two atoms approach each other and their electron clouds overlap and electron density is greatest between the nuclei. This region of high electron density (the covalent bond) attracts each nucleus and therefore keeps the atoms together.

- Covalent bonding is a strong electrostatic attraction between the nuclei of the bonded atoms and the **shared pair of electrons** between them.
- The distance between the two nuclei is the **bond length**. It is the separation at which the energy of the system is at its lowest.

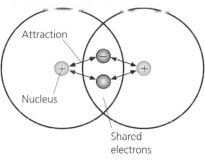

A covalent bond is a strong attraction between the nuclei and the shared pair of electrons between them

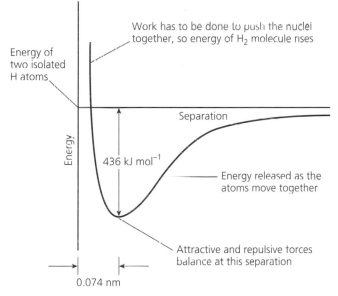

The bond length in a hydrogen molecule is 0.074 nm

Dative covalent bonds are formed when both the shared electrons come from just one of the atoms.

For example, aluminium chloride, $AlCl_3$, will form dimers (a combination of two identical molecules) of Al_2Cl_6. The aluminium atom in aluminium chloride is electron-deficient (only six electrons in its outer shell) but by forming dative covalent bonds the **octet rule** is fulfilled.

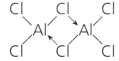

Dative covalent bonding in Al_2Cl_6. The arrows show the electron pairs that have come from chlorine atoms.

Watch out!

Covalent bonds and *dative* covalent bonds are exactly the same – once they have been formed.

Atoms can share more than one pair of electrons and form double or triple covalent bonds:

- a **double bond** results from two shared electron pairs – e.g. in oxygen, O_2, $O=O$
- a **triple bond** results from three shared electron pairs – e.g. in nitrogen, N_2, $N≡N$.

The evidence for covalent bonds

The physical properties of **giant atomic structures** such as diamond provide evidence for the strong electrostatic attraction in covalent bonding. Giant atomic structures are also known as **giant molecular structures**. They are very hard and have high melting temperatures. The covalent bonds are very strong, holding the atoms in place and require a lot of energy to break them before the atoms can move in a liquid.

Electron density maps show high electron density between atoms that are covalently bonded.

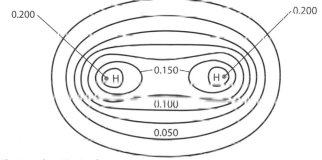

Electron densities in electrons per cubic atomic length

Electron density in the covalent bond between two hydrogen atoms, showing region of high electron density between the two atoms

Covalent dot-and-cross diagrams

Only electrons in the outer shell are shown in these diagrams:
- electron pairs are drawn at four points
- electrons are paired between atoms to form outer shells of eight electrons (two for hydrogen)
- the electrons from one type of atom are usually shown with crosses and those from the other atom with dots.

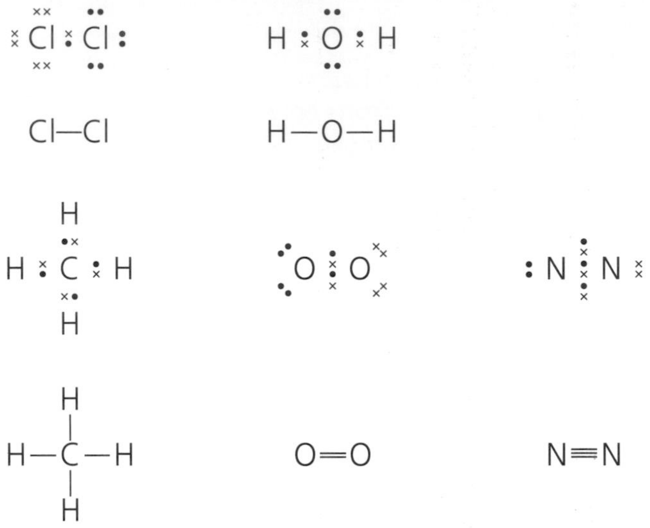

$$\overset{\times\times}{\underset{\times\times}{\times}} Cl \overset{\bullet\bullet}{\underset{\bullet\bullet}{\bullet}} Cl \overset{\bullet\bullet}{\underset{\bullet\bullet}{\bullet}} \qquad H \overset{\bullet\bullet}{\underset{\bullet\bullet}{\bullet}} O \overset{}{\times} H$$

$$Cl-Cl \qquad\qquad H-O-H$$

Dot-and-cross diagrams for some covalent molecules

Lone pairs

In some molecules, not all the electrons in the outer shell may be involved in bonding.
- A non-bonding pair of electrons is called a **lone pair**.
- Lone pairs are shown on dot-and-cross diagrams on one atom only (not shared).
- Lone pairs affect the shape of molecules (see Unit 2, page 73).

Worked Example

Draw a diagram to show the bonding in silane, SiH_4.

This is simply asking you to draw the dot-and-cross diagram.

Follow the stages shown in the diagram.

When you have finished, check that each atom has the right number of its own outer shell electrons (4 for Si and 1 for each H) and a total of 8 electrons surrounding the Si, with 2 for each H.

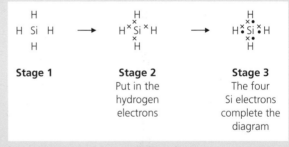

Stage 1

Stage 2
Put in the hydrogen electrons

Stage 3
The four Si electrons complete the diagram

Drawing the dot-and-cross diagram for SiH_4

Quick Questions

1. Why does a covalent bond hold a pair of atoms together?
2. Draw diagrams to show the outer electrons in Cl_2O, CO and Al_2Cl_6.
3. How do the physical properties of diamond give evidence to support the theory of covalent bonding?

Thinking Task

Are all covalent bonds the same?

Metallic bonding

Metals consist of **giant lattices** of metal ions in a sea of **delocalized electrons**. The metal ions vibrate about fixed points in the solid lattice, being held in place by the electrons (also vibrating) around them. It is the **outer electrons** of the metal that have become **delocalized** – they are no longer associated with one particular atom.

Metallic bonding is the strong attraction between metal ions and the sea of delocalized electrons.

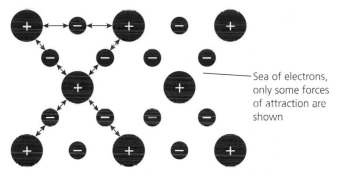

Sea of electrons, only some forces of attraction are shown

The attraction between metal ions and electrons keeps the ions in place

ResultsPlus
Watch out!

Be careful! Use the correct names for particles – *atoms* in *covalent* bonds, but *ions* in *ionic* bonds.

The typical characteristics of metals can be explained using this simple model of metallic bonding.

- *Electrical conductivity* – the delocalized electrons are free to move in the same direction when an electric field is applied to the metal; the movement of charged particles is an electric current
- *Thermal conductivity* – the delocalized electrons transmit kinetic energy (heat) through the metal, from a hot region to a cooler one, by colliding with each other.
- *High melting temperatures* – the positive ions are strongly held together by the attraction of the delocalized electrons; it takes a lot of energy to break the metallic bonds and allow the particles to move around in the liquid state.
- *Malleability and ductility* – metals can be hammered into shape (malleable) or stretched into wire (ductile) because the layers of positive ions can be forced to slide across each other while staying surrounded by the sea of delocalized electrons.

Quick Questions

1   Why do metals generally have high melting temperatures?
2   Explain why metals are good conductors of heat.
3   What keeps the ions in a metal in place?

Thinking Task

Why do Group 1 metals have lower melting temperatures than the transition metals?

Topic 4: Bonding checklist

By the end of this topic you should be able to:

Revision spread	Checkpoints	Specification section	Revised	Practice exam questions
Ionic bonding	Recall and interpret evidence for the existence of ions	1.6.1a	☐	☐
	Describe the formation of ions in terms of electron loss or gain	1.6.1b	☐	☐
	Draw dot-and-cross diagrams to represent electronic configuration of cations and anions	1.6.1c	☐	☐
	Describe ionic crystals as giant lattices of ions	1.6.1d	☐	☐
	Describe ionic bonding as the result of strong net electrostatic attraction between ions	1.6.1e	☐	☐
	Recall the trends in ionic radii down a group and for a set of isoelectronic ions	1.6.1f	☐	☐
Lattice energies and Born–Haber cycles	Recall the stages involved in the formation of a solid ionic crystal from its elements	1.6.1g	☐	☐
	Test the ionic model for ionic bonding by comparison of lattice energies from Born–Haber cycles with values calculated from electrostatic theory	1.6.1h	☐	☐
	Use values calculated for standard enthalpies of formation based on Born–Haber cycles to explain why particular ionic compounds exist	1.6.1l	☐	☐
Testing the ionic model	Explain the term polarization as applied to ions	1.6.1i	☐	☐
	Understand that the polarizing power of a cation depends on its radius and charge, and the polarizability of an anion depends on its size	1.6.1j	☐	☐
	Understand that polarization of anions leads to some covalency in an ionic bond	1.6.1k	☐	☐
	Use values calculated for standard enthalpies of formation based on Born–Haber cycles to explain why particular ionic compounds exist	1.6.1l	☐	☐
Covalent bonding	Understand that covalent bonding is strong and arises from the electrostatic attraction between the nucleus and the electrons which are between the nuclei, based on the evidence from: (i) the physical properties of giant atomic structures (ii) electron density maps for simple molecules	1.6.2a	☐	☐
	Draw dot-and-cross diagrams to represent electronic configurations of simple covalent molecules, including multiple bonds and dative covalent bonds	1.6.2b	☐	☐
Metallic bonding	Understand that metals consist of giant lattices of metal ions in a sea of delocalized electrons	1.6.3a	☐	☐
	Describe metallic bonding as the strong attraction between metal ions and the sea of delocalized electrons Use the models of metallic bonding to interpret simple properties of metals	1.6.3b	☐	☐

ResultsPlus
Build Better Answers

1　Draw a dot-and-cross diagram to show the arrangement of electrons in phosphorus trichloride, PCl_3. You need only show the outer electrons.

(2)

☑ Examiner tip

First find phosphorus in the Periodic Table. It is in Group 5, therefore has 5 outer electrons. Now draw the 3 chlorine atoms around the central phosphorus, placing each Cl at three of the four points of the compass, and between each Cl and the P draw a cross representing a chlorine electron. Complete the chlorine atoms by drawing three pairs of crosses around each one, preferably at three compass points.

Now draw a dot next to each of the crosses between the P and Cl atoms. These represent 3 of the 5 phosphorus outer electrons.

Finally place a pair of dots at the fourth compass point, these are the two remaining phosphorus electrons – a lone pair.

◼ **Basic answer:** Many students can draw the correct number of electrons round each Cl atom. (1) However, they may show the symbols for each atom.

▲ **Excellent answer:** To get the second mark, you need to show that there are three shared pairs and one lone pair of electrons round the P atom.　(1)

Answers missing out the pair of non-bonding electrons for P, or showing 8 electrons but not as three shared pairs plus one lone pair, would not gain credit.

2　Explain why beryllium chloride is covalent but magnesium chloride is ionic.

(3)

☑ Examiner tip

This is all about the polarizing power of the Group 2 cations in relation to the chloride anion.

The charges on the beryllium and magnesium ions are the same because they are in the same group. However, the Be^{2+} ion has one fewer electron shells than the Mg^{2+} ion and is very small.

The answer is:

The chloride ion is moderately polarizable.　(1)

The Be^{2+} ion would therefore be very polarizing and polarize the Cl^- ion to form a covalent bond.　(1)

The Mg^{2+} ion is larger and not sufficiently polarizing to form a covalent bond with Cl^-.　(1)

There is frequent confusion between polarizing power and polarizability. Take care not to get the explanation the wrong way round – e.g. 'the chloride ion polarizes the Be^{2+} ion' is incorrect.

▲ **Excellent answer:** Keep your answers concise (short) and to the point (accurate) but ensure you answer fully. In the question about $BeCl_2$ and $MgCl_2$ you have to say something about all three types of ion and the clue is that there are three marks available.

3　Magnesium iodide is a compound of magnesium. The radius of the magnesium ion is 0.072 nm, whereas the radius of the iodide ion is much larger and is 0.215 nm.

　a　Describe the effect that the magnesium ion has on an iodide ion next to it in the magnesium iodide lattice.

(1)

☑ Examiner tip

The electrons around the iodide ion are drawn towards the magnesium ion/Mg^{2+} polarizes I^- ions.　(1)

Take care not to get confused between polarizing power and polarizability. Marks will also be lost if you refer to atoms instead of cations. you must say the Mg^{2+} *ion* polarizes, not Mg^{2+} polarizes, or that iodine/I_2 is polarized.

　b　What **two** quantities must be known about the ions in a compound in order to calculate a theoretical lattice energy?

(2)

☑ Examiner tip

Radius/size (of ions), (1) charge/charge density　(1)

Again, don't refer to atoms – 'atomic radius' is incorrect here. There are no marks for recalling that Coulomb's law is used, unless you also give the terms in the Coulomb's law formula.

c Suggest how the value of the theoretical lattice energy would compare with the experimental value from a Born–Haber cycle for magnesium iodide. Give a reason for your answer. (2)

Examiner tip

The theoretical lattice energy would be less. (1)

Students have difficulty explaining *why* the theoretical value will be less. The theoretical value assumes a fully ionic model. However, the small, highly charged magnesium ion polarizes the iodide ion giving some covalent bonding character. However, the bonding is not fully covalent – you will lose the mark for saying this. An acceptable answer would be simply 'covalent character', or 'the theoretical value assumes a fully ionic model'.

(Adapted from Edexcel Unit test 4 Q1, June 07)

Practice exam questions

1 Theoretical lattice energies can be calculated from electrostatic theory. Which of the following affects the magnitude of the theoretical lattice energy of an alkali metal hydride, M^+X^-? (1)

   A The first electron affinity of X

   B The first ionization energy of M

   C The enthalpy of atomization of M

   D The radius of the X^- ion

   (From Edexcel Unit test 1 Q2, Jan 09)

2 **a** Name the type of bonding present in magnesium chloride. (1)

   b Draw a diagram (using dots and crosses) to show the bonding in magnesium chloride. Include **all** the electrons in each species and the charges present. (3)

   c State the type of bonding that exists in solid magnesium. (1)

   d Explain fully why the melting temperature of magnesium is higher than that of sodium. (3)

   (Adapted from Edexcel Unit test 1 Q19, Jan 09)

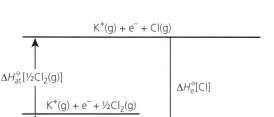
ResultsPlus
Examiner tip

Read the question! This time you are to show all the electrons, not just those in the outer shell. It also reminds you to show the charges – a big hint about the type of bonding – check that your answers match the information given in the questions!

3 **a** Use the following data to complete the Born–Haber cycle for potassium chloride and use it to calculate the electron affinity of chlorine. (5)

	ΔH/ kJ mol^{-1}
1st ionization energy of potassium	+419
Enthalpy of atomization of potassium	+89.2
Enthalpy of atomization of chlorine	+121.7
Enthalpy of formation of KCl(s)	−436.7
Lattice enthalpy of potassium chloride	−711

   b Calcium is in the same period in the Periodic Table as potassium. The lattice enthalpy of calcium chloride is −2258 kJ mol^{-1}. Explain why this is so different from the value for potassium chloride given in **(a)**. (2)

   c Lattice enthalpies may be calculated based on an assumption about the structure of the solid, or found experimentally using data in this Born–Haber cycle.

   The experimentally found lattice enthalpy of potassium chloride is 9 kJ mol^{-1} more exothermic than that calculated; for calcium chloride the experimental value is 35 kJ mol^{-1} more exothermic than that calculated.

   Suggest why the calculated and experimental values are different in both compounds. (3)

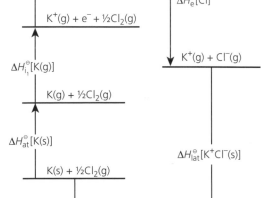

Hazard and risk in organic chemistry

The difference between hazard and risk

Hazard is the potential of a substance or activity to do harm. This is absolute – for example, petrol is toxic and flammable, these properties have been measured and tested and are constant.

Risk is the chance that a substance or activity will cause harm. This is variable – for example, filling a car with petrol carries a low risk of harm due to the safety precautions which are in place, but lighting a fire using petrol is very dangerous (*do not do it!*) since the invisible petrol vapour is not under control.

Types of hazard

Organic chemicals present a range of different hazards. Their containers, transport boxes, storage bottles, lab bottles etc. carry **hazard symbols**. Different hazards require different precautions to be taken:

- **flammable** materials need to be kept away from ignition sources
- **corrosive** materials should be prevented from getting on your skin
- **toxic** materials must not be allowed into the body either through skin absorption or inhalation
- **oxidizing** materials have to be carefully used with fuels or reducing agents.

Reducing risk

The total elimination of risk, while continuing with the normal functioning of society, is almost impossible – e.g. crossing a road without risk is impossible.

A **risk assessment** identifies the **hazards** involved in an activity. It is used to reduce the **risk** of harm to as low as reasonably possible. Any activity or reaction undertaken in a lab requires a risk assessment to have been done before it is carried out. A risk assessment will involve the following steps:

1 Identifying the hazards associated with the chemicals and procedures involved

2 Quantifying the risks by considering, for example, the amount of substances to be used, where the procedure will be done and the experience of the people involved.

3 Identifying who is at risk.

4 Stating the controls used to minimise the risk – e.g. use a fume cupboard.

5 Making a decision as to whether the risk is at an acceptable level for the activity to proceed.

ResultsPlus
Watch out!

A risk assessment does not reduce risk – it *identifies* hazard. Applying risk assessment to an activity reduces the risk of harm to anyone.

Risk can be reduced by:
- Using less material – the reaction is easier to contain and to control, and the risk of spillage is reduced.
- Using lower concentrations of solutions – diluted **corrosive** solutions can become **irritants**, still a hazard but a much reduced one.
- Using an electric heating mantle instead of the naked flame of a Bunsen burner to heat flammable liquids or solutions.
- Using specific protective clothing – e.g. gloves when handling corrosive liquids.
- Doing a reaction in a fume cupboard – thus removing harmful vapours from the work area.
- Reducing the temperature at which the procedure is carried out – thus slowing the reaction and reducing the risk of overheating and too many fumes being produced.
- Changing the materials used – less hazardous compounds may not react as quickly or give as much product, but they will still allow the same reaction to be studied.
- Having assessed the hazards and applied risk assessments, laboratories are very safe places to work – often safer than everyday life!

Worked Example

How would you reduce the risks to a class of Year 10 students investigating the rate of reaction between sodium thiosulfate solution and hydrochloric acid? The equation is:

$$Na_2S_2O_3(aq) + 2HCl(aq) \rightarrow 2NaCl(aq) + SO_2(g) + S(s) + H_2O(l)$$

Work through the risk assessment step-by-step:

1 Identify the hazards associated with the chemicals and procedures involved
   - Refer to *Hazcards* to find the hazards associated with sodium thiosulfate solution, hydrochloric acid and sulfur(IV) oxide.
   - Does the fine suspension of sulfur present a hazard?
   - Is the reaction to be done at various temperatures?

2 Quantify the risks by considering:
   - the volume and concentrations of solutions to be used – will the volume of SO_2 produced exceed the limit for the size of laboratory?
   - therefore, does the reaction have to be done in a fume cupboard?
   - the maximum temperature to be used – would a lower temperature work?
   - if the teacher is aware of the hazards involved?
   - are the students sufficiently skilled and well behaved?
   - are only those in the teaching lab at risk?

4 Minimise the risk by using:
   - for example, $20 \, cm^3$ $0.5 \, mol \, dm^{-3}$ HCl instead of $5 \, cm^3$ $2 \, mol \, dm^{-3}$ acid
   - disposing of the mixture in the fume cupboard sink – thus removing SO_2 fumes from the lab
   - ensuring that any asthmatic students or staff are identified and have their relief medication with them, or remove them from the vicinity.

5 Decide whether the risk is at an acceptable level for the activity to proceed.

Thinking Task

Will the reactions of ^{24}Mg and ^{23}Mg with oxygen be any different? If so, how? If not, why?

Quick Questions

1 What is the difference between a risky experiment and a hazardous chemical?

2 How can the hazard of $2 \, mol \, dm^{-3}$ acid ('corrosive') be reduced to 'irritant'?

3 If a reaction demanded the use of bromine, $Br_2(l)$, (corrosive, very toxic) and not bromine water, how would you reduce the risk to an acceptable level?

Organic compounds and functional groups

Carbon is unique in its ability to form covalent bonds with itself and other non-metals in the same compound — it can make four covalent bonds. Carbon atoms also form strong bonds with hydrogen atoms. Single, double or triple bonds change the arrangement around the carbon atom. All these factors lead to carbon forming millions of **organic compounds**.

Organic compounds form groups (or families), called **homologous series**, in which the compounds have physical properties that are similar, but change gradually with an increase in the number of carbon atoms. For example, the melting temperatures and viscosity of alkanes increase with the number of carbon atoms in their molecules.

Each homologous series is characterised by a general formula and one or more **functional groups** that determine the properties of that series.

carbon atom (2,4)

methane C(2,8) same electronic arrangement as neon

In methane, carbon forms a stable octet by sharing electrons with four hydrogen atoms

Homologous series	General formula	Functional group	Example: structural formula	Example: displayed formula
Alkanes	C_nH_{2n+2}	—C—H	Ethane CH_3CH_3	H—C—C—H (with H atoms)
Alkenes	C_nH_{2n}	C=C	Ethene CH_2CH_2	H₂C=CH₂ displayed
Alcohols	$C_nH_{2n+2}O$	—O—H	Ethanol C_2H_5OH	H—C—C—O—H displayed
Halogenoalkanes	$C_nH_{2n-1}X$	—X	Chloromethane CH_3Cl	H—C—Cl displayed

Homologous series of organic compounds met in AS chemistry

- The general formula of **alkanes** is C_nH_{2n+2} – for example, ethane, C_2H_6 and heptane, C_7H_{16}
- The general formula of **alkenes** is C_nH_{2n} – for example, ethene, C_2H_4 and heptene, C_7H_{14}
- The general formula of **alcohols** is $C_nH_{2n+1}OH$ – for example, ethanol, C_2H_5OH and heptanol, $C_7H_{15}OH$
- The general formula of **halogenoalkanes** is $C_nH_{2n+1}X$ – for example, fluoroethane, C_2H_5F and bromoheptane, $C_7H_{15}Br$.

ResultsPlus
Watch out!

When drawing displayed formulae it is important to remember *all* the H atoms and carefully draw the bonds accurately between bonded atoms, and not somewhere vaguely in their vicinity.

ResultsPlus
Examiner tip

Make sure you are clear when you are writing about alkanes and alkenes. It is far too easy for the middle 'a' or 'e' to become neither. The examiner needs to be able to read what you mean – you cannot rely on 'benefit of the doubt' here because the difference between the two words is significant.

Quick Questions

1 What are the formulae for:
   a hexane   **b** octene
   c pentanol   **d** iodopropane?
2 What is the functional group in the alcohols?
3 What stays the same within an homologous series?

Thinking Task

Carbon and silicon are both in Group 4. Why, then, is the C—C bond so much stronger (347 kJ mol⁻¹) and therefore more stable than the Si—Si bond (+226 kJ mol⁻¹)?

Naming and drawing organic compounds

Prefix	Number of carbon atoms in the main carbon chain
meth-	1
eth-	2
prop-	3
but-	4
pent-	5
hex-	6
hept-	7
oct-	8
dec-	10

The first part of the organic name gives the longest carbon chain in the molecule

The International Union of Pure and Applied Chemistry (IUPAC) have a system of rules, which, when followed to the letter, keep the matter of nomenclature (naming compounds) quite straightforward.

The **systematic name** of an organic compound gives:
- the number of carbon atoms in the molecule
- whether the molecule is a straight or branched chain or a ring
- the homologous series to which it belongs
- the names of any other atoms in the compound.

The first part (**prefix**) of the name gives the number of carbon atoms in the longest chain – e.g. *meth*ane, CH_4, has 1 carbon atom and *pent*ane has 5.

The second part (**suffix**) of the name gives the type of compound – the homologous series to which it belongs. For example, methan*ol*, CH_3OH, is an alcoh*ol*, pent*ene* is an alk*ene* and ethan*al* is an *al*dehyde.

Suffix	Homologous series
-ane	alk**ane**
-ene	alk**ene**
-ol	alcoh**ol**
-oic	carboxyl**ic** acid
-al	**al**dehyde
-one	ket**one**

The last part of the name gives the type of compound

Organic compound names start looking complicated when we have to locate branches, **functional groups** and carbon–carbon multiple bonds. We have to put numbers into the names:
- The numbers refer to the carbon atoms in the longest chain to which any functional group (including multiple bonds and side-chains) is attached.
- The longest carbon chain is numbered from the end that will give the lowest numbers.

Look at the molecules in the diagram. Make sure you know why they are called but-1-ene and 3-ethylhexane.

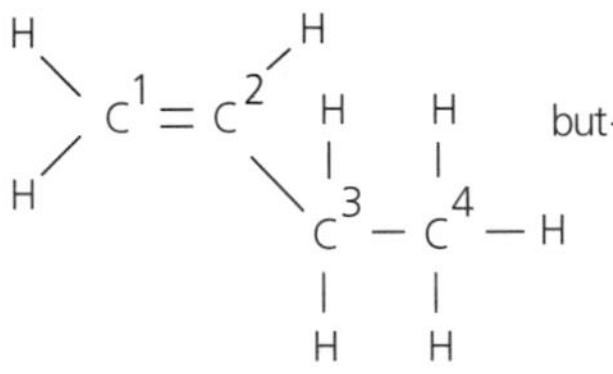

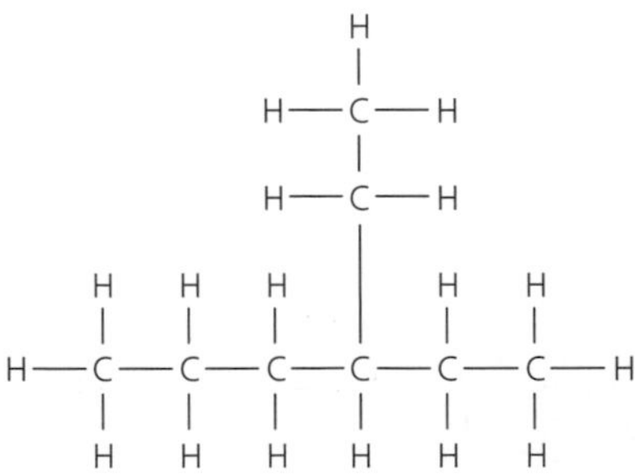

ResultsPlus
Examiner tip

Don't be fooled by a long chain with a right-angle in the displayed formula – too many students are! Count the carbon atoms through each 'route' along the chain and then check it again.
Look carefully at the worked examples.

Carbon chains are numbered from the end that will produce the lowest numbers

Branches are often **alkyl** groups. These are alkane molecules that have lost a hydrogen atom – e.g. the ethyl group comes from ethane.

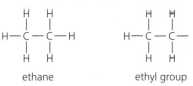

ethane ethyl group

The ethyl group is an ethane molecule that has lost a hydrogen atom

The first three alkyl groups are:

- methyl CH_3—
- ethyl C_2H_5— or CH_3CH_2—
- propyl C_3H_7— or $CH_3CH_2CH_2$—

Multiple side chains or functional groups are listed in alphabetical order. If two functional groups are attached to the same carbon atom then the number of that atom is repeated. If repeated functional groups are the same then the name contains *di,* which becomes *tri* for three and *tetra* for four.

Look at the examples on this page and, using the rules, make sure you can produce the same names!

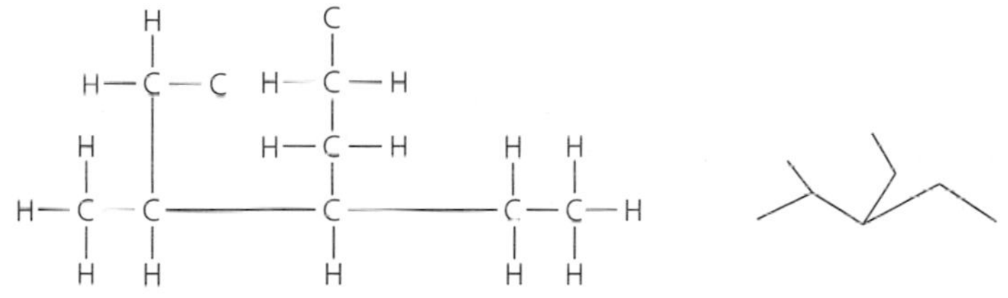

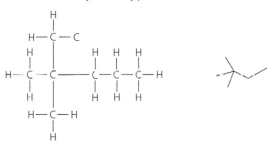

3-ethyl-2-methylpentane

2,2 dimethylpentane

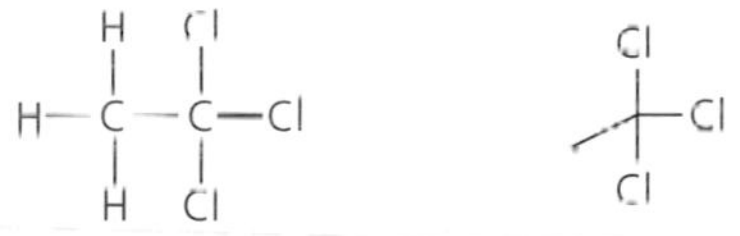

1,1,1-trichloroethane

Examples of naming organic compounds using the IUPAC nomenclature rules

ResultsPlus
Watch out!

Be careful with the commas and hyphens in these names – commas separate numbers, hyphens separate words from numbers. Here are some more examples:

- 2-methylpropan-1-ol
- 1,1-dibromoethane
- 2,3-dibromo-1-chloro-2-fluorobutane

This last example reminds you to list functional groups alphabetically.

A **skeletal formula** is a simplified organic formula. It shows the bonds in the carbon skeleton and any functional groups, with hydrogen atoms removed from alkyl chains. You need to be able to use skeletal formulae, so these have been included with the displayed formulae for you to recognise. The ends of the lines or junctions represent carbon atoms, and hydrogen atoms take up any unused carbon bonds.

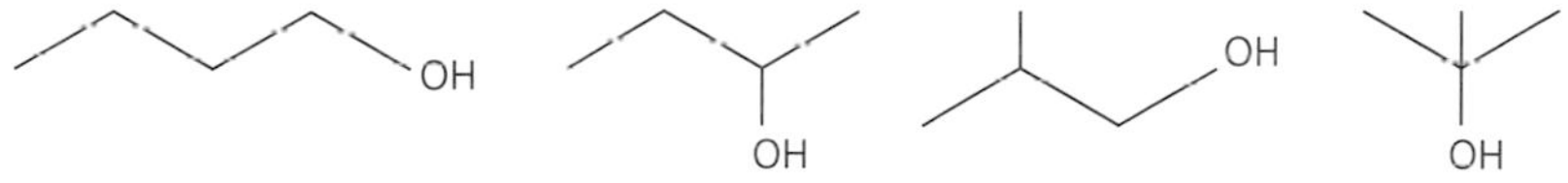

Skeletal formulae of the four structural isomers of butanol

Quick Questions

1 What is the structural formula of ethane-1,2-diol?
2 What is the IUPAC systematic name for carbon tetrachloride?

Thinking Task

What would the skeletal formula of cyclohexene look like?

Hydrocarbons: alkanes and alkenes

Alkanes

Alkanes are **saturated** hydrocarbons – they contain the maximum number of hydrogen atoms, which means single bonds *only*. The general formula of the alkanes is C_nH_{2n+2} – the alkanes form an **homologous series**.

Alkanes with the *same* molecular formula but *different* structures are **structural isomers**. The number of structural isomers increases rapidly as the number of carbon atoms increases – methane, ethane and propane have no structural isomers, butane has two and pentane has three.

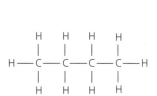

butane:
boiling temperature −0.4°C

2-methylpropane:
boiling temperature −11.6°C

The two isomers of C_4H_{10}

Hydrocarbon fuels and alternative fuels

Alkanes are used as **fuels**. They burn in excess oxygen to produce carbon dioxide and water only. For example:

methane: $CH_4(g) + 2O_2(g) \rightarrow CO_2(g) + 2H_2O(g)$

'petrol': $CH_3CH(CH_3)CH(CH_3)CH(CH_3)CH_3(l) + 12\frac{1}{2}O_2(g) \rightarrow 8CO_2(g) + 9H_2O(g)$

- Alkanes are produced by the fractional distillation of crude oil. Fractions are mixtures of hydrocarbons.
- Kerosene and diesel oil can be cracked to yield smaller, more useful molecules – e.g. kerosene will crack to give gasoline.
- **Reforming** is similar to **cracking**. It breaks up heavy straight-chain molecules into smaller, branched molecules but also increases the proportion of aromatic compounds in the mixture.

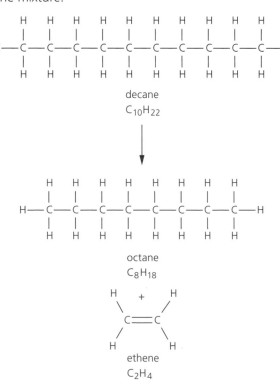

decane
$C_{10}H_{22}$

octane
C_8H_{18}

ethene
C_2H_4

Cracking produces useful, smaller alkanes and alkenes

- The crude oil from which fuels can be derived is not a **sustainable** resource.
- Increased numbers of vehicles are increasing the amount of carbon dioxide emissions from burning traditional fuels.
- This is thought to be causing **global warming**.

Alternative fuels are being developed to try to reduce greenhouse gas emissions and to increase sustainability.
- **Hydrogen** has been considered as an alternative fuel to hydrocarbons. It burns to make water only – this is very 'clean', but H_2O is also a greenhouse gas!
- Hydrogen also combines with oxygen in fuel cells to generate electricity, again producing only water.
- However, the energy used in producing hydrogen (from electrolysis of brine or reaction of methane with steam) and pressurizing the hydrogen comes from power plants that produce greenhouse gases such as carbon dioxide.
- The carbon in methane is also converted to carbon dioxide during the manufacture of hydrogen – hydrogen may not be as 'carbon neutral' as people like to think.

Biofuels, such as bioethanol and biodiesel, are becoming more available as alternatives to fossil fuels. The plants that are used to make them absorb carbon dioxide from the atmosphere as they grow and then release it when the biofuel is burned. Not accounting for the energy requirements of the manufacturing process, biofuels are more carbon neutral than coal, oil and gas.

Alkenes

Alkenes are **unsaturated** hydrocarbons – they contain carbon–carbon **double bonds** (C=C). The general formula of the alkenes with one C=C bond is C_nH_{2n}. The alkenes form an **homologous series**.

The C=C bond consists of two parts: a σ **(sigma) bond** and a π **(pi) bond**:
- In sigma bonds the electron cloud is concentrated between the two nuclei.
- In pi bonds the electron cloud is not in a line between the two nuclei, but above and below the plane of the molecule.
- Atoms bonded by a σ bond can rotate about its axis, but the π bond does not allow rotation and thus affects the structure and reactions of alkenes.

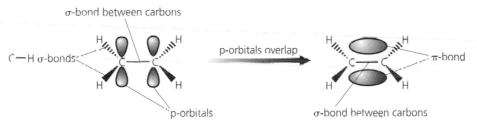

A single bond is a σ bond, a double bond has both a σ bond and a π bond

Quick Questions

1 Draw the structural and skeletal formula for the isomers of pentane, and name them.
2 What does distillation do to crude oil?
3 How does catalytic cracking help to satisfy the demand for motor fuel?

Thinking Task

Is hydrogen the fuel of the future? What are the advantages and disadvantages of using hydrogen to power our modern lives?

Naming geometric isomers

Geometric isomers occur due to the lack of rotation around a C=C double bond. Different groups can therefore be arranged on different sides of the molecule. Geometric isomers:
- often have different properties – e.g. melting and boiling temperatures
- can have different chemical properties, but more often they are the same.

Cis–trans isomerism

The traditional system of naming these geometric isomers is based on **cis–trans isomerism**:
- *cis*-isomers have the functional groups on the same side of the double bond
- *trans*-isomers have the functional groups on opposite sides of the double bond.

The *cis* and *trans* isomers of but-2-ene are shown below.

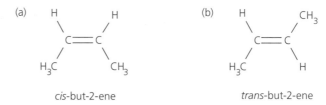

Cis- and trans-but-2-ene

E–Z isomerism

The *cis–trans* system does not work for classifying all geometric isomers. **E–Z isomerism** is the new IUPAC system for naming these compounds – it works for all geometric isomers. The *E–Z* system assigns ranks to the groups attached to the C=C bond. These ranks are based on atomic number – the higher the atomic number, the higher the rank.

> **Worked Example**
>
> What are the *E–Z* names of the isomers of butane?
>
> ---
>
> Look at the isomers of but-2-ene. Consider each carbon in the double bond – they both have a carbon and a hydrogen attached. Carbon has the higher atomic number and thus ranks higher than hydrogen.
> - When the highest-ranking groups on each carbon are on the same side of the double bond, they are *together* (*zusammen*) – that is the Z-isomer.
> - When the highest-ranking groups on each carbon are on opposite sides of the double bond, they are *apart* (*entgegen*) – that is the E-isomer.
>
> Hence:
> - isomer **(a)** is *cis*-but-2-ene or Z-but-2-ene
> - isomer **(b)** is *trans*-but-2-ene or E-but-2-ene.
>
>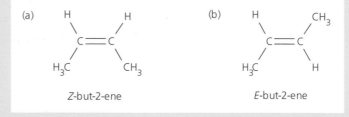
>
> *Z-but-2-ene and E-but-2-ene*

ResultsPlus
Watch out!

Cis- is not always *Z-* and *trans-* is not always *E-*. You have to work it out each time.

E–Z isomers? Just follow these steps:
Step 1: Rank the groups attached to each carbon by atomic number.
Step 2: High ranks on the **s**ame side = Z.
Step 3: High ranks on opposit**e** sides = E.

Worked Example

Look at the isomers of 1-bromo-2-chloro-2-fluoro-1-iodoethene and give them their *E*- and *Z*- names.

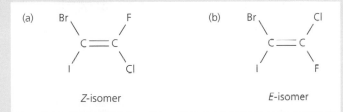

Geometric isomers of 1-bromo-2-chloro-2-fluoro-1-iodoethene

Step 1: Ranking the groups attached to each carbon we get iodine higher than bromine, chlorine higher than fluorine.

Step 2: In isomer **(a)** the high ranks (I and Cl) are on the same side, therefore it is the *Z*-isomer.

Step 3: In isomer **(b)** the high ranks (I and Cl) are on opposite sides, therefore it is the *E*-isomer.

Geometric isomerism occurs widely in food products, both natural and synthetic. Careful analysis of food demands the ability to be able to distinguish between the various geometric isomers, whether classified using the *cis–trans* or the *E–Z* system.

Quick Questions

1 Name the isomers in this diagram. Give both *cis–trans* and *E–Z* names if they exist.

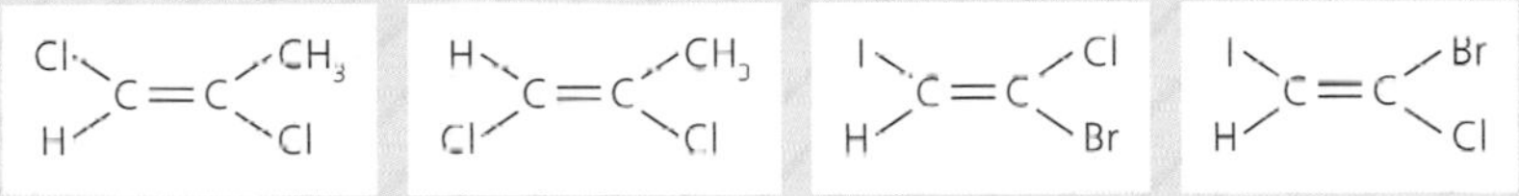

2 In what ways do geometric isomers often differ?

Thinking Task

When does the *cis–trans* system not work?

Reactions of alkanes

Combustion

Alkanes burn in excess oxygen to produce carbon dioxide and water only – for example:
- methane, the main constituent of natural gas, used to heat many homes

$$CH_4(g) + 2O_2(g) \rightarrow CO_2(g) + 2H_2O(g)$$

- octane, present in motor fuel, burns according to

$$C_8H_{18}(l) + 12\tfrac{1}{2}O_2(g) \rightarrow 8CO_2(g) + 9H_2O(g)$$

When burned in plenty of oxygen, alkanes are clean fuels. When burned in a limited supply of oxygen, alkanes undergo *incomplete* combustion and form carbon monoxide. This is very dangerous – it is gaseous, odourless, invisible and toxic – it can be fatal.

Incomplete combustion of alkanes can occur in:
- poorly ventilated rooms

$$CH_4(g) + 1\tfrac{1}{2}O_2(g) \rightarrow CO(g) + 2H_2O(g)$$

- and in car engines

$$C_8H_{18}(l) + 8\tfrac{1}{2}O_2(g) \rightarrow 8CO(g) + 9H_2O(g)$$

ResultsPlus
Watch out!

The symbol Cl· shows the unpaired electron in a chlorine atom, emphasising that it is a radical – it *is not* an ion.

Substitution

A **substitution reaction** is a replacement reaction. For example, hydrogen in an alkane can be replaced by a different atom or group. Chlorine will substitute hydrogen in methane to form a halogenoalkane:

$$CH_4(g) + Cl_2(g) \rightarrow CH_3Cl(g) + HCl(g)$$

This is called **free-radical substitution**. A **radical** is a species with an unpaired electron, such as Cl· formed from the dissociation of chlorine:

$$Cl_2 \rightarrow Cl· + Cl·$$

Alkanes undergo substitution by halogens at about 300°C or in ultraviolet light. The substitution reaction happens in steps. The sequence of steps is shown in a **reaction mechanism**.

Mechanism of free-radical substitution by chlorine

Step 1: **Initiation** – ultraviolet light provides the energy to break the Cl—Cl bond, generating two **free radicals**. The electron pair in the covalent bond is split up and each chlorine atom takes one electron with it. A chlorine radical is simply a single chlorine atom, with extra energy:

$$Cl—Cl \xrightarrow{UV} Cl\cdot + Cl\cdot$$

Initiation will keep going so long as there are chlorine molecules in the gas mixture.

(a) $Cl \overset{h\nu}{\underset{\wedge}{\textstyle =}} Cl \longrightarrow Cl\bullet + Cl\bullet$

(b) $\overset{\frown}{Cl}\bullet\bullet\overset{\frown}{Cl} \longrightarrow Cl\bullet + Cl\bullet$

Mechanism for formation of chlorine free radicals. (a) Curly arrows show the movement of electrons; (b) Half-headed curly arrows show the movement of an electron

Step 2: **Propagation** – there are two reactions in this step:

$$CH_4 + Cl\cdot \rightarrow \cdot CH_3 + HCl$$
$$\cdot CH_3 + Cl_2 \rightarrow CH_3Cl + Cl\cdot$$

Propagation maintains the concentration of chlorine-free radicals in the gas mixture – they can react again in many propagation stages until the termination step has removed all the radicals.

Step 3: **Termination** – there can be several ways in which free-radical reactions are stopped; they all involve the combination of two radicals:

$$Cl\cdot + Cl\cdot \rightarrow Cl_2$$
$$\cdot CH_3 + \cdot CH_3 \rightarrow C_2H_6$$
$$\cdot CH_3 + Cl\cdot \rightarrow CH_3Cl$$

Termination removes radicals from the gas mixture, thus stopping the reaction.

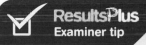

ResultsPlus
Examiner tip

Mechanisms can look daunting to learn – but they are quite easy really. Mechanisms are all about movement of electrons, and electrons move from centres of negative charge to centres of positive charge. At each step identify the negative and positive areas and put in a curly arrow to show the electron movement. (Half-headed arrows show one electron moving, full arrows represent two electrons moving.) And that's it! In this reaction there's only one step requiring curly arrows.

ResultsPlus
Watch out!

Where you draw curly arrows in mechanisms is very important. Firstly, they show where electrons move from and where they move to. Look at the diagrams on this page. The arrows start where the electrons are initially. The arrow heads show where the electrons go to – usually between two atoms.
Secondly, too many students lose marks in mechanisms due to careless, vague and ambiguous curly arrows. Make sure yours are spot on!

ResultsPlus
Examiner tip

Learn this mechanism! Just six reactions – one in the 1st step, two in the 2nd step and three in the 3rd step.

Quick Questions

1   What is a radical?
2   Why is energy required to initiate the substitution of chlorine in methane?
3   Explain why chloromethane is not the only possible product of this reaction.
4   Why is incomplete combustion of hydrocarbon fuels dangerous?

Thinking Task

How would an excess of chlorine produce CCl_4 in the halogenation of methane?

Reactions of alkenes

The C=C double bond in alkenes makes them much more reactive than the alkanes. Alkenes react across the double bond – an **addition reaction** with another substance to form a single, saturated product. The reactions you have to know are:

- alkene + hydrogen → alkane
- alkene + halogen → halogenoalkane
- alkene + hydrogen halide → halogenoalkane
- alkene + acidified $KMnO_4$ → a diol

Addition of hydrogen

This is called **catalytic hydrogenation**. It is used in the manufacture of margarine to turn unsaturated vegetable oils into a soft, saturated fat. It has to be done at around 200°C in the presence of a finely divided (high surface area) **nickel catalyst**.

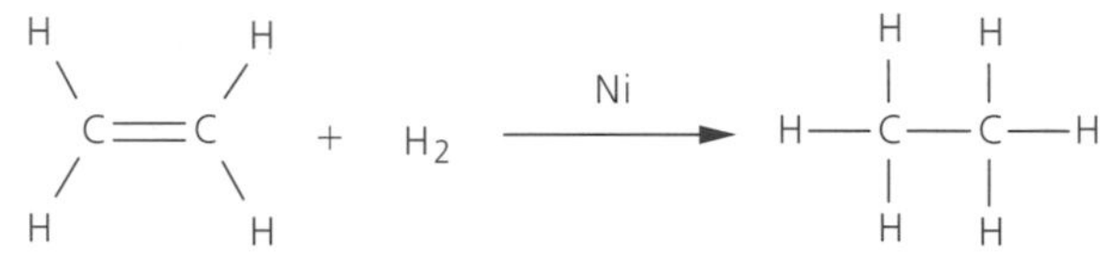

Catalytic hydrogenation of ethene produces ethane

> **ResultsPlus Examiner tip**
>
> Shaking a sample of a compound with a little bromine water is the test for the presence of C=C. When asked to describe the colour changes seen in a reaction, always state the colour it starts and the colour it becomes. Here, bromine is decolorised – it starts brown and becomes colourless.

Addition of halogens

Ethene reacts with bromine or bromine water to produce the **halogenoalkane** called 1,2-dibromoethane, CH_2BrCH_2Br. This is a disubstituted halogenoalkane. The bromine is decolorised in the reaction and the product is a colourless liquid.

The other halogens also react with alkenes to produce disubstituted halogenoalkanes. This is an addition reaction and the mechanism is described below.

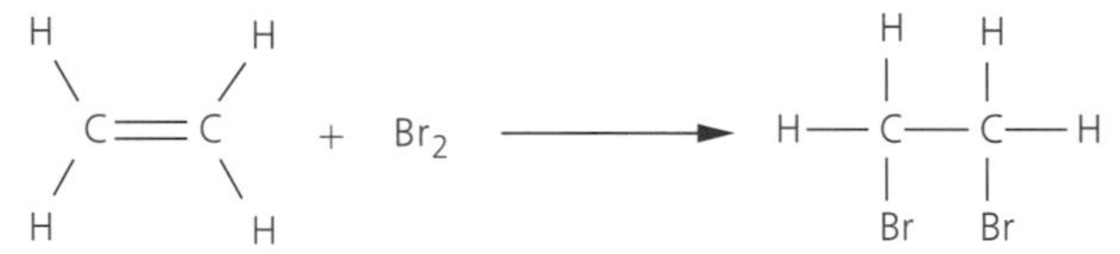

1,2-dibromoethane

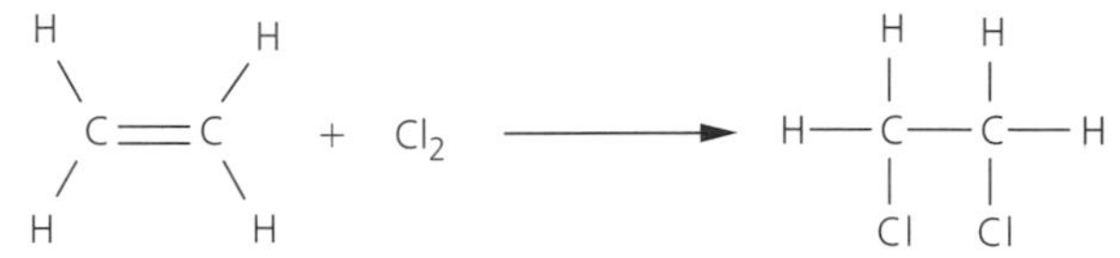

1,2-dichloroethane

Bromination is a typical halogenation reaction – addition across the double bond

Addition of hydrogen halides

Ethene reacts with hydrogen bromide to produce the halogenoalkane bromoethane, CH_3CH_2Br. This is a monosubstituted **halogenoalkane**.

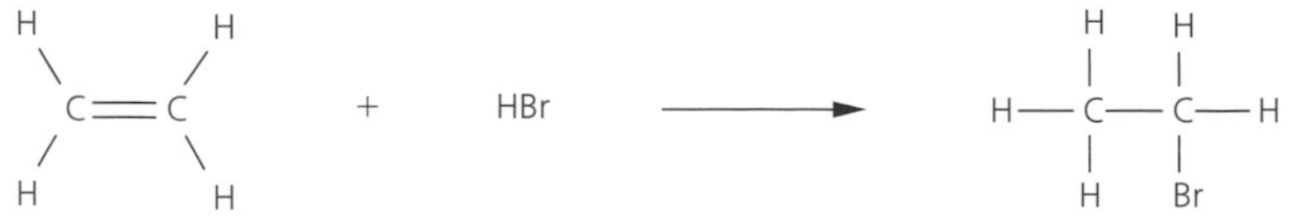

Ethene Bromoethane

Hydrogen bromide adds across the double bond in alkenes

Reaction with acidified potassium manganate(VII)

Potassium manganate(VII) ($KMnO_4$) is acidified with dilute sulfuric acid. The reaction results in oxidation of the alkene to form a diol (with two —OH groups). The potassium manganate(VII) is decolorised – it changes from purple to colourless. This reaction can also be used as a test for alkenes.

ethane-1,2-diol

Acidified manganate(VII) is decolorised as it oxidizes alkenes

Addition of bromine – the mechanism

Addition across the C=C double bond is a two-step mechanism:
- electrophilic attack on the double bond
- nucleophilic attack on the resulting **carbocation** – an ion in which the positive charge is carried on a carbon atom is called a carbocation.

Overall, the mechanism is called **electrophilic addition** (see diagram).
- When bromine approaches the double bond in the ethene, its electron cloud shifts due to the repulsion from the high electron density – this shift produces an instantaneous **dipole**.
- The polarized bromine acts as an **electrophile** and attracts an electron pair from the double bond to form a covalent C—Br bond, a positive carbocation and a bromide ion.
- The bromide ion (a **nucleophile**) now attacks the carbocation – this is a nucleophilic attack.

Electrophilic addition of bromine to ethene – the curly arrows show where the electrons start and where they go, in this case between atoms to form a bond

Addition of hydrogen bromide – the mechanism

The HBr molecule has a permanent dipole. HBr acts as **electrophile** and attracts an electron pair from the double bond. Therefore, the mechanism is the same as that for bromination of ethene.

Hydrogen bromide adds across the double bond in an alkene – the first step is electrophilic, the second step is nucleophilic

Addition of hydrogen bromide to propene

This is also an **electrophilic addition** reaction, but because propene is asymmetrical there are two possible products.

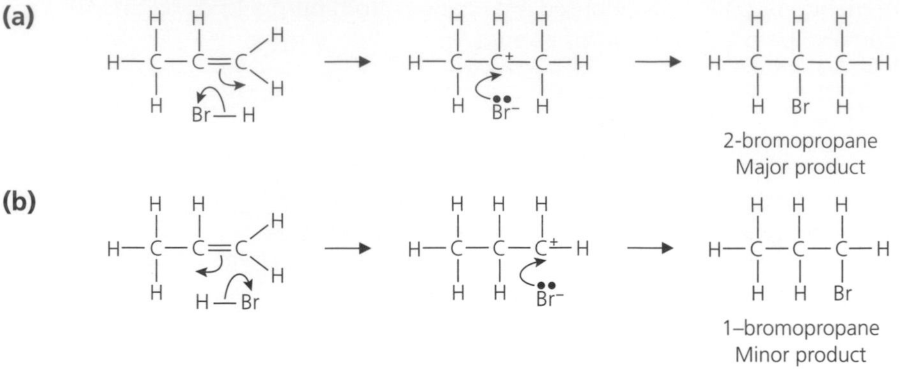

(a)

2-bromopropane
Major product

(b)

1–bromopropane
Minor product

Electrophilic addition of hydrogen bromide to propene

- The carbocation formed in **(a)** is more stable because the methyl groups on either side donate electron density and stabilise the positive charge.
- Therefore the carbocation formed in **(a)** is longer lived (more stable), than that in **(b)** – hence 2-bromopropane dominates the product mixture.

Evidence for the reaction mechanisms

If these addition reactions are done in the presence of competing nucleophiles, such as chloride ions, then a mixture of two products is formed, as shown below.

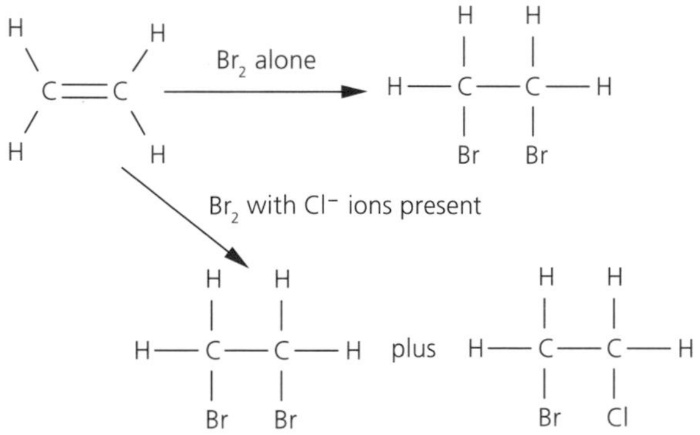

Having formed a carbocation, Cl⁻ ions compete with Br⁻ ions resulting in a product mixture

These two products can only be formed if the first step results in the formation of a carbocation, to which a negative ion can bond.

Bromine water and the test for a double bond (C=C)

- If a compound contains at least one C=C double bond it will decolorise bromine water.
- The reaction is an addition reaction, similar to the addition of bromine to an alkene.
- However, in aqueous bromine, both Br^- from the bromine and OH^- from the water can add to the carbocation.
- Two different reactions can take place (see diagram).

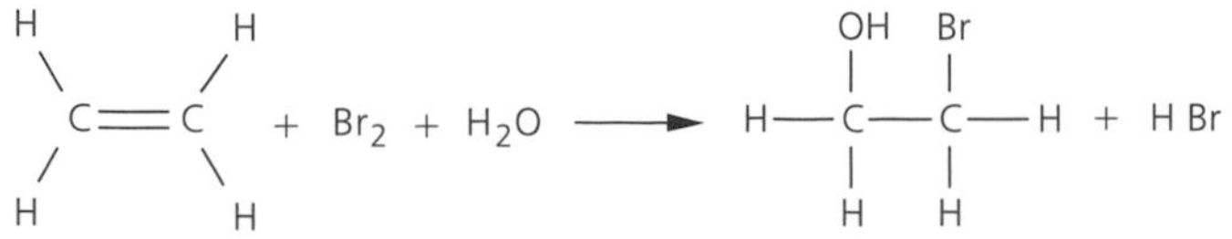

The test for C=C in alkenes. The double bond reacts with water and bromine.

ResultsPlus
Examiner tip ☑

Draw your mechanisms very carefully and you can't go wrong! Remember – electrons are attracted to centres of positive charge. So the first arrow you draw should be from a region of high electron density (a negative ion or a double bond) to a more positive area. That will push electron charge away, so the second arrow should show that. Practise, over and over again, each mechanism, until you can draw them confidently.

⚙ Thinking Task

Why are potassium manganate(VII), bromine and bromine water decolorised when they react with alkenes?

❓ Quick Questions

1 Under what conditions will hydrogen react with ethene?
2 What is produced by reacting ethene with chlorine?
3 What is an electrophile?

Polymers

Addition polymerisation

A **polymer** is formed when a very large number of **monomers** join together to form a chain. A **monomer** is a small molecule. Alkenes undergo **addition polymerisation** to form a poly(alkene) – the reaction occurs by addition across the double bond.

A **repeat unit** is the monomer with the double bond replaced by a single bond and two side links drawn. Many polymer formulae include the letter n to represent the number of monomer molecules linked – in the case of some polymers n will be around 10 000 to 20 000 or bigger!

Ethene polymerizes to form poly(ethene) – three monomers are shown joined as three repeat units

Chloroethene   Poly(chloroethene) – PVC

'n' chloroethene monomers join to form a long chain of 'n' poly(chloroethene) repeat units

A monomer alkene can be identified from the polymer – it includes a $C=C$ bond and the four groups attached to the two carbon atoms.

The polymer lifecycle

Polymers are made from chemicals derived from crude oil, therefore they:
• have high-energy production costs
• use up non-renewable resources.

Disposal of synthetic polymers is a huge problem as they are non-biodegradable (taking up landfill space long-term) and burn to produce toxic gases.

Using renewable or alternative energy resources in the manufacture of polymers reduces the effect on the environment and conserves fossil fuels.

The benefits of recycling polymers are reduction in the:
• consumption of finite oil resources
• need for disposal of plastic
• carbon footprint (mass of carbon dioxide produced in a process)
• energy consumed in manufacture of the product
• consumption of water during manufacture.

But recycling has its own energy costs:
• transport of the material to processing plants
• energy consumed in sorting, melting or shredding
• transport of the new products.

Energy recovery is a new approach to reducing the energy consumption of polymer manufacture. Energy recovery can generate electricity and produce hot water from the high-temperature **incineration** of plastic waste. The very high temperature ensures that toxic gases, such as dioxins, are not produced. This use of plastic waste reduces the amount of fossil fuel used to generate power and heat.

A **lifecycle analysis** quantifies the energy and materials used and any environmental emissions produced during the extraction of raw materials, original manufacture, recycling, reuse or disposal of a polymer product.

ResultsPlus
Examiner tip

Be careful drawing the displayed formulae of polymers – many students find it difficult to draw an unsaturated polymer from alkenes larger than ethene. If you draw the monomer so that the double bond forms the cross-bar of an 'H' then the repeat unit follows exactly the same pattern as in poly(ethene) and poly(chloroethene).

Quick Questions

1   Suggest why high-temperature incineration of plastic waste may not be considered to be sustainable.
2   Draw three repeat units from a section of poly(propene), the addition polymer of propene.
3   What information is required to make a lifecycle analysis for a product?

Topic 5: Introductory organic chemistry checklist

By the end of this topic you should be able to:

Revision spread	Checkpoints	Specification section	Revised	Practice exam questions
Hazard and risk in organic chemistry	Appreciate the difference between hazard and risk	1.7.1c	☐	☐
	Understand the hazards associated with organic compounds and why risk assessments are necessary; suggest ways by which risks can be reduced and reactions can be carried out safely	1.7.1d	☐	☐
Organic compounds and functional groups	Understand that the series of organic compounds are characterised by a general formula and functional groups	1.7.1a	☐	☐
	State the general formula of alkanes	1.7.2a	☐	☐
	State the general formula of alkenes	1.7.3a	☐	☐
Naming and drawing organic compounds	Use IUPAC rules to name compounds relevant to this specification and draw these compounds, using structural, displayed and skeletal formulae	1.7.1b	☐	☐
Hydrocarbons: alkanes and alkenes	State the general formula of alkanes and understand that they are saturated hydrocarbons that contain single bonds only	1.7.2a	☐	☐
	Explain the existence of structural isomers using alkanes (up to C_5) as examples	1.7.2b	☐	☐
	Know that alkanes are used as fuels and are obtained from the fractional distillation, cracking and reformation of crude oil	1.7.2c	☐	☐
	Discuss the reasons for developing alternative fuels in terms of sustainability, reducing emissions and climate change	1.7.2d	☐	☐
	State the general formula of alkenes and understand that they are unsaturated hydrocarbons with a carbon–carbon double bond which consists of a σ and a π bond	1.7.3a	☐	☐
Naming geometric isomers	Explain *E–Z* isomerism (geometric/*cis–trans* isomerism)	1.7.3b	☐	☐
	Understand the *E–Z* naming system and why it is necessary to use this when the *cis-* and *trans-* naming system breaks down	1.7.3c	☐	☐
Reactions of alkanes	Describe the combustion and substitution reactions of alkanes, showing the mechanism of free-radical substitution using curly half-arrows	1.7.2e	☐	☐
Reactions of alkenes	Describe the addition reactions of alkenes: (i) the addition of hydrogen (ii) the addition of halogens (iii) the addition of hydrogen halides (iv) oxidation of the double bond by potassium manganate(VII)	1.7.3d	☐	☐
	Describe the mechanism of: (i) the electrophilic addition of bromine and hydrogen bromide to ethane (ii) the electrophilic addition of hydrogen bromide to propene	1.7.3e	☐	☐
	Describe the test for the presence of C＝C using bromine water and understand that the product is the addition of OH and Br	1.7.3f	☐	☐
Polymers	Describe the addition polymerisation of alkenes and identify the repeat unit given the monomer, and vice versa	1.7.3g	☐	☐
	Interpret information about the uses of energy and resources to show how the use of renewable resources, recycling and energy recovery can contribute to the more sustainable use of materials	1.7.3h	☐	☐

ResultsPlus
Build Better Answers

1 What is the systematic name for the following compound? (1)

$$H_3C-C=C-C-C_2H_5$$ (with CH₃ below)

✓ Examiner tip

The longest carbon chain is 6 carbons long. There is a C=C bond in the chain. The compound is a hexene.

The C=C is on carbon number 2, keeping the number as low as possible. The compound is now a hex-2-ene.

There is a methyl group attached to carbon number 4. The compound is an isomer of 4-methylhex-2-ene.

Consider each carbon in the double bond. They both have a carbon and a hydrogen atom attached.

Carbon has the higher atomic number and thus ranks higher than hydrogen. The highest ranking groups on each carbon are on opposite sides of the double bond, they are *apart*, so it is the *E* isomer.

Hence the compound is *E*-4-methylhex-2-ene. (1)

(From Edexcel Unit test 1 Q17, Jan 09)

2 The mechanism of the reaction represented by the equation:

$$C_2H_4(g) + Br_2(g) \rightarrow CH_2BrCH_2Br(l)$$

is an example of:
  A Free-radical substitution   **B** Free-radical addition   **C** Electrophilic substitution   **D** Electrophilic addition   (1)

✓ Examiner tip

Start by eliminating the impossible answers! If this were a substitution then there would be an extra product that has been substituted. Here there is not, so it cannot be **A** or **C**.

Since there is only one product, the two reactants must have simply been added together – confirming it to be some sort of addition reaction.

Free radical or electrophilic? The only free-radical reaction you have come across in AS chemistry is substitution with alkanes. C_2H_4 is an alkene so this is not a free-radical reaction within your experience. Furthermore, alkenes are reactive – the π bond breaks quite readily and is a region of high electron density. Ethene will be attacked by electrophiles, of which bromine is a common example.

It must be electrophilic addition, answer **D**. (1)

(From Edexcel Unit test 1 Q16, Jan 09)

3 a Name each of the following organic compounds and the homologous series to which it belongs. (4)

✓ Examiner tip

Propan-2-ol (1), an alcohol   (1)

3-methylpentane (1), an alkane.   (1)

  b i Draw the structural formulae, showing all bonds, of the two straight-chain alcohols with molecular formula $C_4H_{10}O$. (2)
    ii Give the systematic names of these compounds. (2)

Examiner tip

i Four carbon atoms so the 'but-' chain should be drawn. An alcohol so there is an —OH group in each.

Draw a 4-carbon chain with an —OH group at the end.   (1)

For the second isomer, draw the —OH attached to a different carbon.   (1)

Some students fail to show the H atoms at the end of the bonds. A stick is not enough – you must show the atoms each bond is connected to.

ii The names are butan-1-ol (1) and butan-2-ol.   (1)

c Pent-2-ene shows geometric isomerism.
  i Draw the structures of its two geometric isomers.   (2)
  ii Explain why geometric isomers can occur.   (1)

Examiner tip

i First, work out the structure of pent-2-ene:

5 carbons in a chain, with a $C = C$ bond on carbon number 2.

The rest of the structure is two H atoms, a methyl group CH_3, and a CH_2CH_3 group.

The E and Z isomers have the H atoms on each carbon either on the same side or on different sides of the molecule.

Take care to read the question. Many students draw the *structural* isomers of pent-2-ene, i.e. pent-1-ene and pent-3-ene.

ii The pent-2-ene molecule has different atoms/functional groups on both (π bonded) C atoms, (1) so the π bond restricts rotation/no rotation about the $C = C$ bond.   (1)

Examiner tip

Relatively few students can explain geometric isomerism clearly. The best answers mention both the restricted rotation and also that the groups on each carbon atom are not the same. Note that the groups do not have to all be different.

(Adapted from Edexcel Unit test 2 Q3, June 07 and Unit test 2 Q1, Jan 07)

Practice exam questions

1 a Draw the structural formula of the substance produced by reacting ethene, C_2H_4, with bromine.   (2)

b The reaction of an alkene with bromine is used as a test for alkenes. Describe in detail how you would carry out this test and what you would see if the test proved to be positive.   (3)

c Ethene can be converted to a polymer called poly(ethene). Draw the structure of part of the poly(ethene) molecule showing the repeat unit clearly.   (2)

2 a Chloroethane is one of the products formed when ethane reacts with chlorine in the presence of ultraviolet light.
  i Name the type of mechanism involved in this reaction.   (1)
  ii Give the mechanism for the initiation step of this reaction.   (2)
  iii Give the equations for the first and second propagation steps.   (2)
  iv Give the equation for a termination step other than that in which chlorine is reformed.   (1)

b Ethene reacts with bromine without the presence of ultraviolet light.
  i Name the type of mechanism involved in this reaction.   (1)
  ii Give the equation for this reaction.   (1)
  iii Describe what you would see in this reaction.   (2)

Unit 1: Practice unit test

Section A

1 Which equation represents the reaction for which the enthalpy change is the lattice energy of sodium fluoride, NaF?

A $Na(s) + \frac{1}{2}F_2(g) \rightarrow NaF(s)$ **B** $Na(g) + F(g) \rightarrow NaF(s)$

C $Na^+(g) + F^-(g) \rightarrow NaF(s)$ **D** $Na(g) + \frac{1}{2}F_2(g) \rightarrow NaF(s)$ (1)

2 The standard enthalpy changes of combustion of carbon, hydrogen and methane are shown in the table below.

Substance	Standard enthalpy change of combustion/kJ mol^{-1}
Carbon, C(s)	−394
Hydrogen, $H_2(g)$	−286
Methane, $CH_4(g)$	−891

Which of the following expressions gives the correct value for the standard enthalpy change of formation of methane in kJ mol^{-1}?

$$C(s) + 2H_2(g) \rightarrow CH_4(g)$$

A $394 + (2 \times 286) - 891$ **B** $-394 - (2 \times 286) + 891$

C $394 + 286 - 891$ **D** $-394 - 286 + 891$ (1)

3 Given the following data:

$$\Delta H_f^\circ [FeO(s)] = -270 \text{ kJ mol}^{-1}$$
$$\Delta H_f^\ominus [Fe_2O_3(s)] = -820 \text{ kJ mol}^{-1}$$

select the expression which gives the enthalpy change, in kJ mol^{-1}, for the reaction:

$$2FeO(s) + \tfrac{1}{2}O_2(g) \rightarrow Fe_2O_3(s)$$

A $(-820 \times \frac{1}{2}) + 270$ **B** $(+820 \times \frac{1}{2}) - 270$

C $-820 + (270 \times 2)$ **D** $+820 - (270 \times 2)$ (1)

4 An organic compound contains 38.4% carbon, 4.80% hydrogen and 56.8% chlorine by mass. What is the empirical formula of the compound?

A C_2H_3Cl **B** CH_3Cl

C C_2H_5Cl **D** $C_3H_5Cl_3$ (1)

5 Which of the following contains the largest number of hydrogen atoms?

A 2 moles of water, H_2O **B** 1.5 moles of ammonia, NH_3

C 1 mole of hydrogen gas, H_2 **D** 0.5 moles of methane, CH_4 (1)

6 Magnesium oxide reacts with dilute hydrochloric acid according to the following equation:

$$MgO(s) + 2HCl(aq) \rightarrow MgCl_2(aq) + H_2O(l)$$

How many moles of magnesium oxide, MgO, are required to neutralize 20 cm^3 of 0.50 mol dm^{-3} hydrochloric acid, HCl?

A 0.0010   **B** 0.0050

C 0.010   **D** 0.020   (1)

7 Hydrogen peroxide decomposes on heating as follows:

$$2H_2O_2(aq) \rightarrow 2H_2O(l) + O_2(g)$$

What mass of hydrogen peroxide is required to give 16 g of oxygen gas?

  A 8.5 g   **B** 17 g

  C 34 g   **D** 68 g   (1)

8 The equation for the dehydration of cyclohexanol, $C_6H_{11}OH$, to cyclohexene, C_6H_{10} is:

$$C_6H_{11}OH \rightarrow C_6H_{10} + H_2O$$

50.0 g of cyclohexanol produced 32.8 g of cyclohexene. Calculate the percentage yield of cyclohexene.

[Molar masses/g mol^{-1}: cyclohexanol = 100; cyclohexene = 82]

  A 32.8%   **B** 40.0%

  C 65.6%   **D** 80.0%   (1)

9 How many isomers are there of C_5H_{12}?

  A two   **B** three

  C four   **D** five   (1)

10 In a molecule of ethene, C_2H_4, how many π (pi) bonds are present?

  A one   **B** two

  C three   **D** four   (1)

11 Propene reacts with hydrogen chloride gas to give mainly

  A 1-chloropropane ($CH_3CH_2CH_2Cl$)

  B 2-chloropropane ($CH_3CHClCH_3$)

  C 3-chloroprop-1-ene ($CH_2{=}CHCH_2Cl$)

  D 1,2-dichloropropane ($CH_3CHClCH_2Cl$)   (1)

Section B

12 a Complete the electronic configurations of magnesium and chlorine atoms.   (2)

         1s   2s   2p   3s   3p

Mg   [$\uparrow\downarrow$] [$\uparrow\downarrow$] [ ][ ][ ]   [ ]   [ ][ ][ ]   (1)

         1s   2s   2p   3s   3p

Cl   [$\uparrow\downarrow$] [$\uparrow\downarrow$] [ ][ ][ ]   [ ]   [ ][ ][ ]   (1)

  b Write the equation, including state symbols, for the reaction of magnesium with chlorine.   (2)

  c The mass spectrum of a sample of chlorine molecules shows three molecular peaks.

     These are formed from the molecules shown below.

Molecule	Percentage abundance
$^{35}Cl{-}^{35}Cl$	56.25
$^{35}Cl{-}^{37}Cl$	37.50
$^{37}Cl{-}^{37}Cl$	6.25

     Calculate the relative molecular mass of chlorine in this sample   (2)

  d Calculate the volume of 4.73 g of chlorine gas at 100°C.   (2)

     [The molar volume of a gas at 100°C = 30.6 dm³ mol^{-1}]

  e State and explain the type of bond that exists in solid magnesium.   (3)

  f State the type of bond that exists in magnesium chloride.

     Draw a dot-and-cross diagram showing the outer shell electrons.   (3)

(From Edexcel Unit test 1 Q1, Jan 07)

13 a Define the term *first ionization energy*. (3)
    b Write an equation, with state symbols, to illustrate the process occurring
       when the second ionization energy of sodium is measured. (2)
    c Explain why the first ionization energies generally increase across the period
       sodium to argon (Na to Ar). (3)
    d Explain why the first ionization energy of aluminium is less than that of
       magnesium. (2)

14 a Define the term *Avogadro constant*. (2)
    b **Z** is a Group 0 element.
       i   1.907g of **Z** contains 2.87×10^{22} atoms of **Z**.
           Calculate the relative atomic mass of **Z**. (2)
           [Avogadro constant $= 6.02 \times 10^{23} \, mol^{-1}$]
       ii Suggest the identity of **Z**. (1)
    c Potassium superoxide, KO_2, reacts with water as follows:

$$2KO_2(s) + 2H_2O(l) \rightarrow 2KOH(aq) + H_2O_2(l) + O_2(g)$$

       i   Calculate the mass of potassium superoxide needed to produce 3.09 g of
           hydrogen peroxide. (3)
           [Molar mass of potassium superoxide, KO_2: $71 \, g \, mol^{-1}$. Molar mass of
           hydrogen peroxide, H_2O_2: $34 \, g \, mol^{-1}$]
       ii Calculate the volume of oxygen produced from the reaction in (i). (1)
           [Molar volume of oxygen under the conditions of the reaction
           $= 24.0 \, dm^3 \, mol^{-1}$]

(From Edexcel Unit test 1 Q6, June 07)

15 a State the general formula of the alkanes, using the letter n to denote the
       number of carbon atoms in each molecule. (1)
    b Alkanes are used as fuels. In the petrochemical industry, useful hydrocarbons
       are often produced from longer chain molecules. Name the type of reaction
       shown below. (1)

    c By what **type** of formula are the organic molecules in **(b)** represented? (1)
    d Another reaction carried out in industry can be represented as shown below.

       Compound 1   Compound 2

       i   Give the molecular formula of compound 2. (1)
       ii Give the name of compound 2. (1)

(From Edexcel Unit test 1 Q24, Jan 09)

Shapes of molecules and ions

Electron sharing in covalent molecules

There are two reasons why covalent bonds form:
- When electrons are shared, the situation is more **stable** than when there are two separate atoms (four attractive forces between positive nuclei and negative electrons instead of two).
- Many atoms (especially H, C, N, O, F, Cl, Br and I) try to share sufficient electrons to achieve the nearest rare gas outer shell electronic configuration.

The first reason is more important than the second – there are many compounds in which one of the atoms has a partly filled outer shell of electrons, like beryllium in beryllium fluoride. There are even some covalent compounds where an atom may end up with an odd number of outer shell electrons. This is often true of nitrogen in nitrogen oxides.

To work out the bonds present in a molecule it is often helpful to draw a **dot-and-cross diagram** to show the electron sharing (see page 42 in Unit 1).

Worked Example

What is the bonding in ammonia, NH_3?

Step 1: Draw the outer electron arrangement for each atom in the molecule. Nitrogen has five outer shell electrons:

Hydrogen has one electron:

Step 2: For the atom with fewest electrons, pair the electrons with electrons from other atoms:

Notice that nitrogen can only form three bonds, despite having five outer shell electrons. In ammonia the fourth electron pair is not bonded.

Shapes of molecules

The **electron-pair repulsion theory** is a model that predicts the shape of a molecule around a central atom:
- Draw the dot-and-cross diagram using the steps above.
- Count up the number of **electron pairs**, including **bonding pairs** of shared electrons and **non-bonding pairs** of electrons.
- These pairs of electrons repel each other because they have the same (negative) charge. They try to get as far apart as possible to minimise repulsions.

Minimum repulsion of outer electron pairs predicts the following **molecule shapes** and **bond angles**.

ResultsPlus
Watch out!

A common mistake is to say that the atoms repel, or that the bonds repel. It is the *electron pairs* that repel.

Pairs of electrons	Shape	Bond angles
2	Linear	180°
3	Trigonal planar	120°
4	Tetrahedral	109.5°
5	Trigonal bipyramidal	90° and 120°
6	Octahedral	90°

Some examples are shown in the table below.

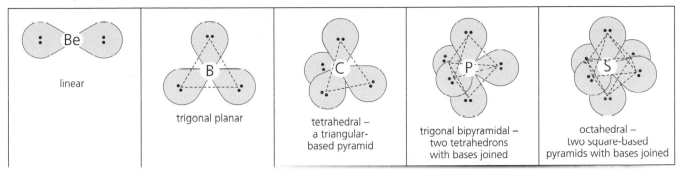

The shapes of molecules can be predicted from the number of electrons around the central atom – they are based on minimising the repulsion between the electron pairs

Non-bonding electrons

Don't forget to show the non-bonding pairs (or **lone pairs**) of electrons in your dot-and-cross diagrams.
- Lone pairs repel more than bonding pairs because they are attracted to a single nucleus and not shared by two atoms.
- Lone-pair repulsion reduces the bond angle between bonding pairs. Each lone pair of electrons reduces the predicted bond angle between bonding electrons by 2.5°.

For example, ammonia has four electron pairs – three bonding pairs and one non-bonding pair. This gives an overall tetrahedral shape for the bond angles. However, the H—N—H bond angle is reduced from 109.5° in a perfect tetrahedral arrangement to 107°, due to the greater repulsion from the lone pair.

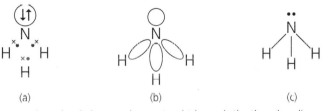

(a)　　　　(b)　　　　(c)

An ammonia molecule has one lone pair, which repels the three bonding pairs

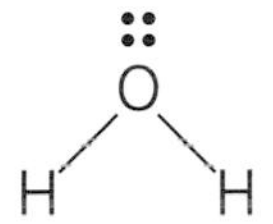

The two lone pairs in a water molecule repel the two bonding pairs. The molecule shape is bent linear.

Water has four pairs of electrons around the oxygen atom – two bonding pairs and two non-bonding pairs. This further reduces the H—O—H bond angle to 104.5°.

Multiple bonds

Electrons in double or triple bonds count as *one* 'pair' for the purpose of determining the shape of the molecule. For example, the C=C bond in ethene counts as one pair, and the C≡C bond in ethyne also counts as one pair.

You also need to remember:
- The double part of the bond is different from the single bond – it may be stronger as in C=O, or weaker as in C=C, than the corresponding single bond. This means there might be slightly more or less repulsion.
- Single bonds are **sigma** (σ) bonds.
- The double part of a bond is a **pi** (π) bond.
- There is no free rotation about double bonds because this would break the π bond.

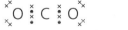　　　　O=C=O

The carbon dioxide molecule has two double bonds and so forms a linear molecule

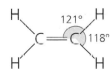　　　

In the ethene molecule, each carbon atom has one double bond and two single bonds – the resulting shape is trigonal planar

Shapes of ions

The ammonium ion has four bonding pairs of electrons round it and so, like methane, its bonds form a tetrahedral shape with a bond angle of 109.5°

The shapes of ions are predicted in the same way as the shapes of molecules. The only thing to be careful of is that each negative charge means that there is an extra electron; and positive ions are short of one electron for each positive charge. In the ammonium ion NH_4^+, the outer shell has 8, not 9 electrons, all of which are bonding. Note that the ammonium ion has **dative covalent bonds**, in which both electrons in the bond come from the same atom.

Carbon structures

You should be familiar with the four **allotropes** of carbon, which have differing molecular structures due to differences in bonding.
- In **diamond**, each carbon atom forms four identical bonds to neighbouring carbon atoms giving a tetrahedral arrangement.
- **Graphite** consists of carbon atoms in layers. Within a layer, each carbon is strongly bonded to three other carbon atoms at 120°. The layers are only weakly bonded to each other. The fourth outer electron from each carbon atom is **delocalized** and free to move, so graphite is a good electrical conductor.
- **Fullerenes** consist of 32 or more carbon atoms – **Buckminsterfullerene** (C_{60}) has 60. Each carbon is bonded to three other carbon atoms to generate a ball-shaped molecule. Like graphite, the fourth outer shell electron is delocalized. However, fullerenes do not conduct electricity because the delocalized electrons cannot move between molecules.
- **Nanotubes** are fullerenes in the form of tubes. Nanotubes are very small, and are stiffer than other known materials. If embedded in polymers they may produce materials with good electrical conductivity and enormous strength.

Quick Questions

1 a Draw a diagram to show the shape of a phosphine molecule, PH_3. Give the bond angle.
  b Explain why a phosphine molecule has the shape you have drawn.
2 a Draw a dot-and-cross diagram to show the arrangement of electrons in an oxonium ion, H_3O^+.
  b Draw a diagram to show the shape of this ion, marking the bond angle clearly.
3 Explain why carbon atoms form four single bonds in a tetrahedral arrangement.

Bond polarity and intermediate bonding

Electronegativity

The electrons shared between atoms in a covalent bond are not necessarily equally shared. The **electronegativity** of an element measures the ability of an atom to attract an electron pair to itself in a covalent bond.

Electronegativity values are highest for the elements in the top right-hand corner of the Periodic Table (ignoring Group 8):
- fluorine has the highest value at 4.0
- oxygen is next at 3.5
- chlorine and nitrogen come next at 3.0

If the difference in electronegativity between two atoms in a molecule is high – roughly above 0.7 and below 1.7 – the **electron density** in the bond is distorted, causing a **dipole**. The resulting bond is **polarized**. In a **polar bond** one end has a slight excess negative charge, written δ–. One end has a slightly positive charge, δ+. 'δ' is the mathematical symbol meaning 'very small'.

Polar molecules

A molecule containing polar bonds is usually polar overall, but sometimes the net effect of several polar bonds cancels out. For example, tetrachloromethane, CCl_4, contains four polar C—Cl bonds. However, the four polar bonds are symmetrically arranged, so the molecule has no net dipole, and it is not polar.

Significance of polar bonds in molecules

Bond polarity is important in determining the reactivity of molecules – particularly in organic chemistry. A polar bond is generally the bond that is likely to break, so this indicates where in a molecule a reaction is likely to occur (see 'Reaction mechanisms' on page 104).

Intermediate bonding

Between the extremes of pure ionic and pure covalent bonding, there is a whole range of **intermediate bonds**. Polarization of ions leads to *distorted* ionic bonds (see page 39). If the polarization is large, the electron density is distorted so much that the ionic bond resembles a covalent bond.

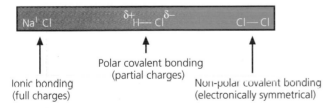

Range of bonding from covalent to ionic

- Ionic bonding (full charges)
- Polar covalent bonding (partial charges)
- Non-polar covalent bonding (electronically symmetrical)

δ+ δ–
H—Cl

For hydrogen chloride, the difference in electronegativity is 3.0 – 2.1 = 0.9, so in the molecule the chlorine is slightly negative and the hydrogen is slightly positive

Chlorine is more electronegative than carbon so each C—Cl bond in tetrachloromethane is polar – however the dipoles act in different directions and cancel out.

ResultsPlus
Watch out!

In exams, many candidates say that in a symmetrical molecule the charges cancel, when they should say that the dipoles cancel.

ResultsPlus
Examiner tip

You should be familiar with the experiment in which a charged rod is brought near a stream of liquid from a burette. If the jet is deflected towards the rod, the liquid contains polar molecules. It doesn't matter if the rod is positively or negatively charged. In liquids the molecules are constantly rotating and will turn so the oppositely charged end is attracted to the rod.

Quick Questions

1 Predict whether the following bonds are polar or non-polar. For each polar bond draw a diagram to show the polarity. State and explain whether the molecule has an overall dipole.
  a C—Cl in chloromethane, CH_3Cl
  b C—Br in tetrabromomethane, CBr_4
  c C—H in methane, CH_4
  d C=O in carbon dioxide, CO_2

Intermolecular forces

Intermolecular forces are:
- forces between *molecules*
- not forces between *atoms* in molecules, which are intramolecular forces (a common student misunderstanding)
- much weaker than covalent bonds.

There are three types of intermolecular force:
- **London forces** – always present
- **Permanent dipole–dipole interactions** – an additional force which may be present
- **Hydrogen bonds** – a further additional force which may also be present.

London forces

These are also known as **van der Waals forces** or dispersion forces. They:
- exist between all molecular substances, and also in monatomic rare gases like helium
- are caused by instantaneous unequal electron distribution in an atom, which gives rise to a temporary or **instantaneous dipole**

This causes an **induced dipole** in the opposite direction on a neighbouring atom, so the two dipoles attract. The induced dipole can cause further induced dipoles on the neighbouring atoms. The net result is a weak attractive force.

There is one key factor that determines the strength of London forces. The more electrons in a molecule, the larger the induced dipoles, and so the larger the London forces. Because molar mass is related to the number of electrons it is a useful guide to the strength of London forces.

Permanent dipole–dipole forces

There is nothing new here – you studied dipoles caused by differences in electronegativity on the previous page! **Permanent dipole–dipole interactions** occur between permanent dipoles in neighbouring polar molecules.
- A polar molecule has two very small opposite charges, $\delta+$ and $\delta-$.
- The $\delta+$ on one molecule then attracts the $\delta-$ on another polar molecule.

Permanent dipole–permanent dipole forces are only really significant if there is significant difference in electro-negativity between the atoms. Hence they are only important in aldehydes, ketones, carboxylic acids, esters and amides.

Hydrogen bonds

Hydrogen bonds are a special type of permanent dipole–dipole force. For hydrogen bonds to form there must be:
- a hydrogen atom bonded *directly* to a highly electronegative atom – fluorine, oxygen or nitrogen
- a lone pair of electrons on the highly electronegative atom.

H—F, H—O and H—N bonds are strongly polar, so there is electrostatic attraction between the $\delta+$ hydrogen atom of one molecule and the $\delta-$ fluorine, oxygen or nitrogen of another molecule.

The $\delta+$ hydrogen atom is within the orbital of the non-bonding pair of electrons on the oxygen atom. It is almost as if the hydrogen has formed a second weak covalent bond – so the hydrogen bond is at 180° to the normal covalent bond between oxygen and hydrogen.

Hydrogen bonding is a very strong interaction. It can be about one tenth of the strength of a covalent bond.

Thinking Task

Hydrogen bond, covalent bond, ionic bond, London forces – are they at all similar? How are they different?

Hydrogen bonding in water – hydrogen bonds are shown between molecules as dashed lines. Note each hydrogen is covalently bonded to an oxygen atom and hydrogen bonded to an oxygen atom in a neighbouring molecule. Each oxygen atom has two covalent bonds and two hydrogen bonds

Explaining trends in physical properties by intermolecular forces

The boiling temperatures of liquids are determined by intermolecular forces. The stronger the intermolecular forces, the higher the boiling temperature.

- The boiling temperature of the alkanes increases as chain length increases. This is because the intermolecular London forces are stronger for larger molecules.
- London forces are stronger between straight-chain alkane molecules than between branched alkane molecules with the same number of carbon atoms. This is because the chains can line up with each other (greater surface area in contact) to maximise the force. This is the reason why 2 methylpropane has a lower boiling temperature than its isomer butane.
- Hydrogen bonding explains why alcohols have much higher boiling temperatures (lower **volatility**) than alkanes of similar molar mass and a similar number of electrons. For example, ethanol, $M_r = 46$, boils at 79°C while butane, $M_r = 44$, boils at −42°C. Both have similar London forces because they have similar numbers of electrons. But ethanol can form additional hydrogen bonds between molecules.

Hydrogen bonding between molecules in ethanol – notice how the hydrogen involved in hydrogen bonding is attached to the oxygen and not a carbon

- The boiling temperature of the hydrogen halides rises from HCl to HI because the London forces are stronger due to increasing numbers of electrons. Hydrogen fluoride is exceptional – it has weaker London forces because it has the fewest electrons. However, it has a much higher boiling temperature than the other hydrogen halides. The reason is that it has additional hydrogen bonds.

Quick Questions

1 a List the following compounds in order of increasing boiling temperature:

   1-bromobutane, 1-bromo-2-methylpropane, butane, 2-methylpropane

   b Explain your answer in terms of intermolecular forces.

2 Why does ethanol, C_2H_5OH, have a much higher boiling temperature than methoxymethane, CH_3OCH_3?

3 Can hydrogen bonds form between:
   a CH_4 and PH_3
   b CH_4 and NH_3
   c CH_4 and HF
   d H_2O and HF?

4 Draw a diagram to show the intermolecular forces between methanol and water molecules. State the type of intermolecular force involved, and give the bond angle between the molecules.

Solubility

You will have done experiments to see which compounds are soluble in different solvents, which liquids mix with each other (dissolve) and which do not. There is a simple general rule in terms of intermolecular forces that *like dissolves like*. This is not a sufficient answer as an *explanation* in an examination, but a useful rule that is quick to apply.

If you are asked to discuss the solubility of one liquid in another, you need to focus on two things:

1 the intermolecular forces in each separate liquid before mixing
2 the potential for new intermolecular forces between the two substances when mixed.

- If the forces of attraction between the molecules of one liquid are stronger than the intermolecular forces between the molecules of the two different liquids, then the liquids do not mix.
- The energy given out when new attractions form between the two liquids will be insufficient to overcome the energy required to break the existing forces.

For example, halogenoalkanes are not soluble in water because there is only weak interaction between water molecules and halogenoalkane molecules (London forces and dipole–dipole interactions). The energy released by these new interactions is not enough to break the strong hydrogen bonds between water molecules.

Worked Example

State and explain the solubility of hexane in water.

Hexane molecules are held to each other by London forces (van der Waals forces). Water molecules are held together by hydrogen bonds. Hexane can't make hydrogen bonds with water, so there is not enough energy to break the strong hydrogen bonds in water – and so the two liquids do not mix or dissolve in each other. They are **immiscible**.

Worked Example

State and explain the solubility of ethanol in water.

Water molecules are bound to each other by hydrogen bonds. Ethanol molecules are also bound to each other by hydrogen bonds, though there are also some weak London forces between the short carbon chains. Ethanol can form hydrogen bonds to water molecules, so they are soluble in each other, or **miscible**.

Note that pentan-1-ol is not soluble in water because the London forces between the long carbon chains are too strong to break, even though the OH groups can form hydrogen bonds with water.

Solubility of ionic compounds in water

Many ionic salts dissolve in water. The energy required to overcome the strong electrostatic forces in the ionic lattice and separate the ions is supplied by the energy released when polar water molecules are attracted to the ions. The positive ions are attracted to the $\delta-$ oxygen atoms in water molecules, and the negative ions are attracted to the $\delta+$ hydrogen atoms. This energy released is called the **hydration energy** and the ions are said to be **hydrated** because they are surrounded by water molecules.

Quick Questions

1 Suggest why many stains cannot be removed by water, but can be removed with ethanol.
2 Suggest why 1,1,1-tetrachloroethane is not miscible with water, but dissolves in hexane.

Topic 1: Bonding and intermolecular forces checklist

By the end of this topic you should be able to:

Revision spread	Checkpoints	Specification section	Revised	Practice exam questions
Shapes of molecules and ions	Use electron pair repulsion theory to predict shapes of simple molecules and ions	2.3a	☐	☐
	Recall and explain the shapes of simple molecules and ions from Units 1 and 2 including $BeCl_2$, BCl_3, CH_4, NH_4^+, H_2O and CO_2	2.3b	☐	☐
	Predict shapes of molecules and ions, similar to those in b above	2.3c	☐	☐
	Predict and understand bond angles and relative bond lengths	2.3d	☐	☐
Bond polarity and intermediate bonding	Explain the meaning of electronegativity	2.4a	☐	☐
	Use electronegativity to explain polar bonds and predict whether compounds will be ionic or covalent	2.4b	☐	☐
	Use knowledge of polar bonds and molecule shape to predict whether a molecule is likely to be polar	2.4c	☐	☐
	Describe and interpret experiments to determine whether molecules in liquids are polar or non-polar	2.4d	☐	☐
Intermolecular forces	Understand the nature of the three types of intermolecular force	2.5a	☐	☐
	Describe the different structures formed by carbon and their applications	2.3e	☐	☐
	Relate the physical properties of materials to intermolecular forces including: (i) the effect of carbon chain length on boiling temperature of alkanes (ii) the effect of branching in carbon chains on boiling temperature of alkanes (iii) the high boiling temperature of alcohols compared to alkanes with similar numbers of electrons (iv) trends in boiling temperatures of hydrogen halides	2.5b	☐	☐
Solubility	Carry out experiments on the solubility of compounds in different solvents Explain choice of solvents and discuss factors which determine solubility including: (i) solubility of ionic compounds in water and hydration of ions; (ii) solubility of alcohols in water through hydrogen bonding; (iii) insolubility of non-polar and polar organic molecules in water, because of their inability to hydrogen bond; solubility of compounds in solvents other than water depending on similarity of intermolecular forces	2.5c 2.5d	☐	☐

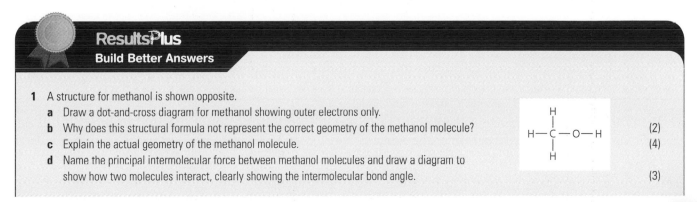

ResultsPlus
Build Better Answers

1 A structure for methanol is shown opposite.
   a Draw a dot-and-cross diagram for methanol showing outer electrons only.
   b Why does this structural formula not represent the correct geometry of the methanol molecule? (2)
   c Explain the actual geometry of the methanol molecule. (4)
   d Name the principal intermolecular force between methanol molecules and draw a diagram to show how two molecules interact, clearly showing the intermolecular bond angle. (3)

Examiner tip

a

b The bond angles around the carbon and oxygen are not 90° but 109.5° for H—C—H (1) and 104.5° for C—O—H (1)
Answer the question as fully as possible – here it is important to give the value of both bond angles.

c The carbon has four pairs of bonding electrons (1) which repel each other and adopt the minimum repulsion tetrahedral arrangement. (1)

The oxygen has two pairs of bonding electrons and two pairs of non-bonding electrons (1). Once again this gives an overall tetrahedral arrangement, but this time the C—O—H bond angle is reduced (1) (because non-bonding pairs of electrons repel more than bonding pairs of electrons).

■ **Basic answer:** In questions like this many candidates do little more than state that the bonds or pairs of bonding electrons repel. This alone would not gain any marks – it is important to state that there is maximum separation or minimum repulsion.

Many candidates also lose marks by not stating the number of bonding pairs around the carbon and the oxygen atom.

▲ **Excellent answer:** The best answers explain why the C—O—H bond angle is reduced compared to the H—C—H bond angle.

It is important to use complete statements – like bonding pairs of electrons. You may be penalised for just saying 'bonding pairs'.

d Hydrogen bonding (1)

H—C—O—H---O—C—H   bond angle O—H---O is 180°

It is fine to draw displayed formulae with apparently incorrect bond angles within the molecule. The question asks only for the *intermolecular* bond angle. Notice the C—H bonds are not involved in hydrogen bonding, and the conventional use of three dashes to represent the hydrogen bond, which places the O, H and O in a straight line.

(Adapted from Chemistry (Nuffield) Unit 2 Q2, June 05)

Practice exam questions

1 Which of the following compounds is likely to be the most soluble in water?
   A PF_3 **B** CH_3CH_2OH **C** $CH_2CH_2CH_3$ **D** CH_4 (1)

2 Which of these four molecules – BF_3, I_2, CO_2 and CF_4 – are not polar?
   A all four **B** CO_2 and I_2 **C** I_2 and CF_4 **D** BF_3 and CO_2 (1)

3 Which one of the following molecules has the most covalent character?
   A LiCl **B** NaCl **C** KCl **D** RbCl (1)

4 Which one of the following molecules does not exhibit hydrogen bonding?
   A H_2O **B** NH_3
   C $CH_3CH_2CH_2CH_2OH$ **D** $CH_3CH_2CH_2OCH_3$ (1)

5 a Boron, nitrogen and oxygen form fluorides with formulae BF_3, NF_3 and OF_2. Draw the shapes you would expect for these molecules, suggesting a value for the bond angle in each case. (6)

   b Another fluoride of nitrogen has the formula N_2F_2.
      i Draw a dot-and-cross diagram for the electronic structure of N_2F_2 showing only the outer electrons. (2)
      ii Draw diagrams to show two possible shapes for the N_2F_2 molecule. (2)
      iii Suggest why the bond energy of the N—F bond in N_2F_2 is significantly different from the N—F bond energy value in NF_3. (1)

   c BF_3 and NF_3 react together readily to give a solid with the formula BF_3NF_3.
      i Draw a dot-and-cross diagram for the electronic structure of BF_3NF_3 showing only the outer electrons. (2)
      ii What is the type of bond between nitrogen and boron atoms? (2)
      (Adapted from Chemistry (Nuffield) C2 Q1, June 96)

Oxidation and reduction

Recall the definitions of oxidation and reduction that you met in your GCSE work.

Oxidation is:
- the addition of oxygen or removal of hydrogen
- an increase in 'positiveness' due to loss of electrons.

Reduction is:
- the removal of oxygen or addition of hydrogen
- a decrease in 'positiveness' due to gain of electrons.

In this section we will refine these terms for describing processes involving **electron transfer**.

Oxidation numbers

It is useful to have a number that tells us 'how oxidized an element is in a compound'. For example, in sodium chloride, sodium has an **oxidation number** of +1, which tells us it has been oxidized by losing one electron. Chlorine has an oxidation number of −1, which indicates that it has been reduced by gaining one electron.

Wider definitions of oxidation and reduction are:
- oxidation happens when an element *increases* its oxidation number in a reaction
- reduction happens when an element *decreases* its oxidation number in a reaction.

Strong oxidizing agents (or **oxidants**) have high oxidation numbers and can oxidize other atoms, molecules or ions by taking electrons away from them. **Reducing agents** can donate electrons and get oxidized.

Writing ionic half-equations

Reduction and oxidation always take place together in a reaction. That is why such reactions are called **redox reactions**. If at least one element changes its oxidation number in a reaction, it is a redox reaction.

It is often helpful to break redox reactions into two ionic **half-equations** – one representing oxidation, and the other reduction. A simple example is burning magnesium:

$$Mg(s) + \tfrac{1}{2}O_2(g) \rightarrow MgO(s)$$

for magnesium: $Mg(s) - 2e^- \rightarrow Mg^{2+}(s)$
for oxygen: $\tfrac{1}{2}O_2(g) + 2e^- \rightarrow O^{2-}(s)$

Notice this makes it quite clear that magnesium is oxidized (electron loss) and oxygen is reduced (electron gain).

Rules for determining oxidation numbers

Oxidation numbers are determined by applying a set of logical rules. You need to learn these very carefully.

1 Uncombined elements have oxidation number zero – for example $H_2(g)$, $Na(s)$.
2 In monatomic ions (single atom ions), the oxidation number is the charge on the ion – so Fe^{2+} has oxidation number +2, O^{2-} has oxidation number −2.
3 In a neutral compound, the sum of the oxidation numbers is equal to zero – for example in NaCl, Na is +1 and Cl is −1, and +1 − 1 = 0.
4 In a polyatomic ion (containing more than one atom), the sum of the oxidation numbers is equal to the charge on the ion – for example SO_4^{2-} has S at +6 and O at −2, and $+6 + (-2 \times 4) = -2$.
5 Many elements almost always have the same oxidation numbers. You need to learn these carefully.

Element	Usual oxidation number	Exceptions
Group 1	+1	
Group 2	+2	
Al	+3	
H	+1	Metal hydrides where it is −1
F	−1	
O	−2	If combined with F, where it is positive (e.g. in OF_2 it is +2), or in peroxides where it is −1 (e.g. H_2O_2)
Cl	−1	If combined with F or O where it is positive
Br	−1	If combined with F, O or Cl where it is positive
I	−1	If combined with F, O, Cl, or Br where it is positive

These follow from each other, the first invariable oxidation number taking priority over the next and so on. For example, sodium is +1, and it can form a hydride with hydrogen. Hydrogen must be −1 in sodium hydride, NaH, by rule 3, because in a compound the sum of the oxidation numbers is zero. Normally, of course, hydrogen is +1.

Worked Example

What is the oxidation number of chromium in the dichromate ion $Cr_2O_7{}^{2-}$?

This is an ion and so the sum of the oxidation states of the atoms of oxygen and chromium in the ion is equal to the charge on the ion. Let the oxidation number of chromium be x:

$$2x + (7 \times -2) = -2$$
$$2x = +12$$

So $x = +6$, so the oxidation number of chromium is +6. Incidentally this confirms that this ion is a powerful oxidizing agent.

Note, the oxidation number of an element in a compound does not mean that the element is in the form of ions with that charge. It would be impossible to get Cr^{6+} ions.

Oxidation numbers in covalent compounds

Though they do not contain ions, oxidation numbers can still be calculated for elements in covalent compounds, and are useful for seeing if a compound has been oxidized.

For example, what happens when methanol, CH_3OH, reacts to form methanal, CH_2O? In both compounds, the hydrogen and oxygen have fixed oxidation numbers, so we can find the oxidation number of carbon by setting up equations, using x to represent the oxidation number of carbon:

for methanol $x + (4 \times +1) + (-2) = 0$ gives $x = -2$
for methanal $x + (2 \times +1) + (-2) = 0$ gives $x = 0$

So we can see that the oxidation number of carbon has increased (from −2 to 0) and become more positive – carbon has been oxidized.

Naming compounds using oxidation numbers

Elements that can have different oxidation numbers can form compounds of the same name with different properties – for example, 'iron chloride' can be $FeCl_2$ or $FeCl_3$. There is a way of naming compounds uniquely by including the oxidation number in Roman numerals after the element – these two iron compounds are called iron(II) chloride and iron(III) chloride, respectively.

This is particularly useful for transition elements, which can have a variable oxidation number.

In oxoacids, the oxidation number of the central atom is given after the rest of the name, which always ends in '-ic'. For example, H_3PO_4 is called phosphoric(V) acid. In the salts of common acids, oxidation states are not needed. For example, Na_2SO_4 is called sodium sulfate, and $NaNO_3$ is sodium nitrate.

Quick Questions

1 Find the oxidation number of nitrogen in each of the following compounds, and state their names: N_2O, HNO_3, $NaNO_2$ and $(NH_4)_2SO_4$.

2 Give the full names of:
 a CuO
 b Cu_2O
 c $Cr_2(SO_4)_3$
 d $CrSO_4$

Redox reactions

Disproportionation reactions

A reaction in which one of the reactants is both oxidized and reduced is called **disproportionation**. For example, in the disproportionation of hydrogen peroxide:

$$2H_2O_2(aq) \rightarrow 2H_2O(l) + O_2(g)$$

In hydrogen peroxide, the oxidation number of oxygen is -1 because it is a 'peroxide'. In water oxygen has its usual oxidation number of -2. The oxygen gas is an element and its oxidation number is 0. So, oxygen is oxidizing (-1 to 0) and reducing (-1 to -2) itself in the same reaction.

Balancing redox equations using oxidation numbers

This is a high-level skill, but you may be asked to balance a complex redox equation. It is also essential in Unit 5!

The total increase in the oxidation number of the element oxidized must balance the total decrease in the oxidation number of the element reduced.

It is best to learn the steps below.

Step 1: Write down the oxidation numbers of the elements that change underneath each element in the unbalanced equation.

Step 2: Calculate the total change in oxidation number for each element – remember that 'change in' means (product value – reactant value).

Step 3: Calculate the electron transfer needed for each element.

Step 4: Balance the number of electrons for each change by multiplying the relevant items by the appropriate number.

Step 5: Balance oxygens by adding water to the side of the equation short of oxygen.

Step 6: Balance hydrogens by adding hydrogen ions ($H^+(aq)$) to the side short of hydrogen.

Step 7: Add up all the charges on each side of the equation to check that they balance.

> **ResultsPlus**
> **Watch out!**
>
> Make sure you have learned the rules for determining oxidation numbers and can write ionic half-equations (see page 77).
> If there are two atoms of an element in the formula, as in I_2, give the oxidation number as 2×0 and make sure you 'pencil in' two corresponding atoms or ions on the other side of the equation.

> **ResultsPlus**
> **Examiner tip**
>
> Some credit may be given in exams for showing the correct oxidation numbers of the elements that change oxidation number clearly.

Worked Example

Balance the equation for the reaction between chlorate(I) ions ClO^- and Fe^{2+} ions to form chloride ions and iron(III) ions Fe^{3+}.

Step 1: $ClO^-(aq) + Fe^{2+}(aq) \rightarrow Cl^-(aq) + Fe^{3+}(aq)$
$\qquad\quad +1 \qquad\quad +2 \qquad\qquad -1 \qquad\quad +3$

Step 2: Fe: $+2 \rightarrow +3$, so the change in ON is $+1$
$\qquad$ Cl: $+1 \rightarrow -1$, so the change in ON is -2

Step 3: Cl: 2 electrons needed, Fe: 1 electron lost

Step 4: The change in ON for iron from Fe^{2+} to Fe^{3+} needs to be multiplied by 2 to balance for electrons:

$$ClO^-(aq) + 2Fe^{2+}(aq) \rightarrow Cl^-(aq) + 2Fe^{3+}(aq)$$

Step 5: Add water to the right-hand side:

$$ClO^-(aq) + 2Fe^{2+}(aq) \rightarrow Cl^-(aq) + 2Fe^{3+}(aq) + H_2O(l)$$

Step 6: Add $H^+(aq)$ to the left-hand side:

$$2H^+(aq) + ClO^-(aq) + 2Fe^{2+}(aq) \rightarrow Cl^-(aq) + 2Fe^{3+}(aq) + H_2O(l)$$

Step 7: Charge check – there are 5 net positive charges on each side of the equation.

Quick Questions

1 Thiosulfate ions react with acid:

$$S_2O_3^{2-}(aq) + 2H^+(aq) \rightarrow S(s) + SO_2(g) + H_2O(l)$$

   a Calculate the oxidation numbers of sulfur throughout this reaction.

   b Use these to explain why this can be classed as a disproportionation reaction.

2 Use oxidation numbers to balance the equation for the reaction between manganate(VII) and iron(II) ions in aqueous solution:

$$MnO_4^-(aq) + Fe^{2+}(aq) \rightarrow Mn^{2+}(aq) + Fe^{3+}(aq)$$

Properties and reactions of Group 2 elements and compounds

Trend in first ionization energy

You are expected to know that **first ionization energy** decreases down Group 2 because:

- Atomic radius is increasing as the additional electrons fill higher energy levels, so the distance between the positive nucleus and the negative electrons is increasing and the force of attraction is less.
- **Shielding** of the nucleus increases. Inner shell electrons shield the two outer electrons from the nucleus, further reducing the force of attraction.

These two factors far outweigh the increase in nuclear charge down the group.

Reactions of Group 2 elements

Reactivity usually increases down the group, as the ionization energy decreases.

- All the Group 2 metals burn in air or oxygen to form solid metal oxides, often burning with a very bright flame. Reactivity increases down the group. In general:

$$Mg(s) + \tfrac{1}{2}O_2(g) \rightarrow MgO(s)$$

For barium:

$$Ba(s) + \tfrac{1}{2}O_2(g) \rightarrow BaO(s)$$

- All the Group 2 metals burn in chlorine to form solid metal chlorides. In general, reactivity increases down the group.

$$Mg(s) + Cl_2(g) \rightarrow MgCl_2(s)$$
$$Ca(s) + Cl_2(g) \rightarrow CaCl_2(s)$$

- Calcium and the elements of the group below it react with water similarly to form hydrogen and the corresponding hydroxide. Reactivity increases down the group.

$$Mg(s) + 2H_2O(g) \rightarrow Mg(OH)_2(aq) + H_2(g)$$
$$Sr(s) + 2H_2O(l) \rightarrow Sr(OH)_2(aq) + H_2(g)$$

Beryllium does not react with water, due to the formation of an insoluble oxide layer. Magnesium reacts very slowly with cold water, but with steam it reacts vigorously to give magnesium oxide, which is almost insoluble.

Reaction of Group 2 oxides

- From calcium down the group, the oxides react with water to form the corresponding hydroxides. For example:

$$CaO(s) + H_2O(l) \rightarrow Ca(OH)_2(aq)$$

Calcium hydroxide has many uses in water purification, and making whitewash, mortar and plaster.

- All the oxides neutralize hydrochloric acid and nitric acid to form the corresponding chlorides or nitrates:

$$MgO(s) + 2HCl(aq) \rightarrow MgCl_2(aq) + H_2O(l)$$
$$CaO(s) + 2HNO_3(aq) \rightarrow Ca(NO_3)_2(aq) + H_2O(l)$$

- The hydroxides of the Group 2 elements react with dilute acids in a similar way:

$$Sr(OH)_2(aq) + 2HCl(aq) \rightarrow SrCl_2(aq) + 2H_2O(l)$$

Solubility of Group 2 hydroxides and sulfates

- The solubility of hydroxides *increases* down the group. Calcium hydroxide is slightly soluble giving a solution called limewater. This solution reacts with carbon dioxide to form a white precipitate of calcium carbonate, $CaCO_3$. On further addition of carbon dioxide, the solution clears because soluble calcium hydrogencarbonate forms, $Ca(HCO_3)_2$.

- The solubility of sulfates *decreases* down the group. Again, calcium sulfate is somewhere in between soluble and insoluble. Barium sulfate is used in 'barium meal' X-ray investigations because it is insoluble, and so it is not absorbed from the gut. It is useful to use barium compounds, despite their poisonous nature, because barium shows up well on X-rays because it has so many electrons.

Thermal stability of Group 1 and 2 carbonates and nitrates

Thermal stability means the stability of a compound when it is heated. The carbonates and nitrates of Groups 1 and 2 follow the same patterns of thermal stability.

- From sodium carbonate down Group 1, the carbonates do not **decompose** on heating.
- From sodium nitrate down Group 1, the nitrates decompose to form the corresponding nitrite and oxygen. For example, for potassium nitrate:

$$KNO_3(s) \rightarrow KNO_2(s) + \tfrac{1}{2}O_2(g)$$

- Lithium carbonate, and all the Group 2 carbonates, decompose on heating to form carbon dioxide and the corresponding oxide. For example:

$$Li_2CO_3(s) \rightarrow Li_2O(s) + CO_2(g)$$
$$CaCO_3(s) \rightarrow CaO(s) + CO_2(g)$$

The second reaction occurs when chalk or limestone is heated to form calcium oxide (also known as 'quicklime').

- Lithium nitrate, and all Group 2 nitrates, decompose on heating to form oxygen, nitrogen dioxide and the corresponding oxide.

Although you do not need to remember the degrees of hydration, given the formula of any nitrate you should be able to write the equation for its thermal decomposition. For example, for magnesium nitrate 6-water:

$$Mg(NO_3)_2 \cdot 6H_2O(s) \rightarrow MgO(s) + \tfrac{1}{2}O_2(g) + 2NO_2(g) + 6H_2O(l)$$

When carried out in the laboratory, you would expect to see:

- the solid dissolving in the water of crystallization
- the solution boiling, and water condensing at the top of the test tube
- a solid reforming as the water boils away
- the solid melting
- and finally a brown gas (nitrogen dioxide) given off, and also a gas that relights a glowing splint.

Going down Group 2, both the carbonates and nitrates become more stable to heat. This is because the **ionic radius** of the positive ion increases down the group and the oxides with smaller ions are more stable, due to polarization of the oxide ions giving additional covalent bonding.

Flame test for Group 1 and 2 ions

You should be able to describe how to carry out a flame test. Nichrome wire is used (an alloy of nickel and chromium) because it is unreactive. *Concentrated* hydrochloric acid is used because chlorides are soluble and volatile:

- Mix the acid and powdered salt and pick up a small sample on the wire.
- Hold the coated wire in a roaring Bunsen flame.

The colours produced are:

- yellow – sodium
- lilac – potassium and caesium
- red – the rest (lithium, calcium, strontium).
- green – barium (Bag!)
- colourless – magnesium

You should also know how these colours arise. Electrons are given energy and excited to higher energy levels. As they fall back to lower energy levels, energy is released in the form of visible light (except in the case of magnesium when the radiation given out is outside the visible region of the electromagnetic spectrum).

ResultsPlus
Examiner tip

In the written and practical tests, you will be expected to interpret the results of tests on Group 2 compounds. The addition of a solution of barium chloride is the test for sulfate ions – the positive result is the formation of a white precipitate of barium sulfate.

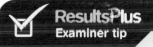

ResultsPlus
Examiner tip

When you carry out the test for oxygen when heating potassium nitrate, it is important not to drop bits of burning splint into the test tube because the mixture may explode.

Quick Questions

1 A white solid reacts with water to form an alkaline solution. This solution reacts with carbon dioxide to form a white precipitate. When a flame test is carried out using this precipitate, a brick-red flame is observed.
 a Identify the white solid, the alkaline solution and the white precipitate.
 b Write equations, with state symbols, for the reactions.

2 Describe how you would carry out a flame test to determine the metal in a sample of rock.

Acid–base titrations

An **acid–base titration** is a technique that involves using accurate volume measurements to find the concentration of acid or alkali solutions.

- The acid, usually hydrochloric acid, of *known concentration*, goes in the **burette**.
- The alkali goes in a conical flask – use a **pipette** and pipette filler to take a *known volume*.
- Always use **methyl orange** indicator, unless you are titrating a weak acid (see below). Methyl orange changes from *yellow* in *alkali* to *red* in *acid* – at the **end-point** it is *orange*.

For a weak acid, such as ethanoic acid, use **phenolphthalein**. This is pink in alkali and colourless in acid – it is very pale pink at the end-point.

Calculating concentrations of unknown solutions

You should be able to calculate the concentration of a solution in $mol\,dm^{-3}$ or $g\,dm^{-3}$ from titration results and the balanced equation for the titration reaction.

From Unit 1 (page 8), to find the number of moles, n, of a solute in a solution use:

$$n = \frac{V \times c}{1000}$$

where V is the volume in cm^3 and c is the concentration in $mol\,dm^{-3}$.

Worked Example

Find the solubility of calcium hydroxide, given that $10\,cm^3$ of a saturated solution of calcium hydroxide was found, by titration, to neutralize $6.10\,cm^3$ of $0.050\,mol\,dm^{-3}$ hydrochloric acid.

Step 1: Find the number of moles of a solute in the hydrochloric acid solution:

$$n = \frac{V \times c}{1000}$$

Number of moles of HCl $= \dfrac{6.10}{1000}\,dm^3 \times 0.050\,mol\,dm^{-3}$

$$= 3.05 \times 10^{-4}\,mol$$

Step 2: From the equation for the reaction:

$$Ca(OH)_2(aq) + 2HCl(aq) \rightarrow CaCl_2(aq) + 2H_2O(l)$$

1 mole of HCl reacts with 0.5 mole of $Ca(OH)_2$.

So the amount of $Ca(OH)_2$ in $10\,cm^3$ of solution

$$= 0.5 \times 3.05 \times 10^{-4}\,mol$$
$$= 1.525 \times 10^{-4}\,mol$$

Step 3: Concentration of $Ca(OH)_2$ solution

$$= \frac{1000}{10} \times 1.525 \times 10^{-4}\,mol\,dm^{-3}$$
$$= 1.53 \times 10^{-2}\,mol\,dm^{-3}$$

Step 4: Solubilities are usually expressed in $g\,dm^{-3}$ so to convert from $mol\,dm^{-3}$ to $g\,dm^{-3}$ multiply by the molar mass of calcium hydroxide, 74.1.

The solubility of calcium hydroxide

$$= 1.53 \times 10^{-2}\,mol\,dm^{-3} \times 74.1\,g\,mol^{-1}$$
$$= 1.13\,g\,dm^{-3}$$

Results of titrations – accuracy, precision and reliability

Results of titration experiments should be recorded in a table like this.

	Titration number		
	1	**2**	**3**
Final burette reading /cm³			
Initial burette reading /cm³			
Titre /cm³			

- The first titre is likely to be inaccurate – a rough titration to get an idea where the end point is – so this result is usually discarded.
- It is usual to record readings to the nearest 0.05 cm³ – this is the limit of **precision** of the burette (the number of significant figures or decimal places that can be read).
- To improve **reliability**, the experiment should be repeated until you have two **concordant** titres, within 0.20 cm³. Then the mean of these two should be found.

When you are carrying out titrations you need to eliminate **systematic errors** as far as possible. Some of these are errors caused by poor technique. You must read the burette at eye level – this means it is unlikely that your first burette reading will be 0.00 cm³, unless you are well over six feet tall! You should also check that the burette jet is filled, and remove the funnel used to fill your burette.

Measurement uncertainties are caused by **random errors**, caused by the limits of accuracy of your apparatus. Repeating an experiment will make the results more reliable, but the uncertainties will remain the same if the same apparatus is used.

Any calculation should be given only to the same number of **significant figures** as the precision of your measurements. You may be penalised in the Unit 3 practical assessment, and in the written exam, for giving an inappropriate number of significant figures for either a mean titre or the calculated solution concentration.

Estimating uncertainties in your final answer

You should be able to estimate the likely **error boundaries** of your calculated results. If you are reading to the nearest 0.05 cm³, each titre will have a total measurement uncertainty of ±0.10 cm³ (there are two readings with an uncertainty of ±0.05 cm³). To calculate the **percentage uncertainty** use:

$$\text{percentage uncertainty} = \frac{\text{uncertainty in the value}}{\text{value}} \times 100\%$$

ResultsPlus
Examiner tip

You should be able to justify your selection of the titrations you use to calculate the mean. If you discard an anomalous result, you should be able to suggest a reason like 'this result was too high because I overshot'.

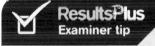

ResultsPlus
Examiner tip

Give your answer to a calculation to the same number of significant figures as the least accurate measurement in the experiment.

Quick Questions

1   Calculate the concentration of a solution of strontium hydroxide, given that exactly 10.0 cm³ of a saturated solution of it reacts with 20.0 cm³ of 0.0500 mol dm⁻³ hydrochloric acid.
2   A student obtained two titres of 7.65 and 7.75 cm³. Calculate the percentage uncertainty in the mean titre.
3   A student obtained the following burette readings in a titration:

	Titration number		
	1	**2**	**3**
Burette reading (final) /cm³	24.05	23.80	24.25
Burette reading (initial) /cm³	0.15	0.00	0.15

   a   State which titres you would use to calculate the mean titre, and give a reason for your answer.
   b   Suggest possible reasons why the reading you have discarded could be inaccurate.

Topic 2: Inorganic Chemistry 1: Group 2 and titrations checklist

By the end of this topic you should be able to:

Revision spread	Checkpoints	Specification section	Revised	Practice exam questions
Oxidation and reduction	(i) Know the rules for working out oxidation numbers (ii) Link oxidation and reduction to electron transfer (iii) Link oxidation and reduction to increase and decrease in oxidation number	2.6a (i), (ii) and (iii)	☐	☐
Redox reactions	(iv) Describe what is meant by a disproportionation reaction	2.6a (iv)	☐	☐
	Write ionic half-equations and use them to write full ionic equations for redox reactions Balance redox equations using oxidation numbers	2.6b	☐	☐
Properties and reactions of group 2 elements and compounds	Explain the trend in ionization energies down Group 2	2.7.1a	☐	☐
	Recall the reactions of Group 2 elements with oxygen, chlorine and water	2.7.1b	☐	☐
	Recall the reactions of their oxides with water and dilute acid, and the hydroxides with dilute acid	2.7.1c	☐	☐
	Recall the trends in solubility of the Group 2 hydroxides and sulfates	2.7.1d	☐	☐
	Recall and be able to explain the trends in thermal stability of nitrates and carbonates of both Groups 1 and 2	2.7.1e	☐	☐
	Recall the characteristic flame colours formed by Group 1 and 2 ions and explain how the flame colours arise	2.7.1f	☐	☐
	Describe and carry out: (i) experiments on the thermal decomposition of Group 1 and 2 nitrates and carbonates (ii) flame tests	2.7.1g (i) and (ii)	☐	☐
Acid–base titrations	Describe and carry out: (iii) acid–base titrations, including knowledge of indicators and appropriate colour changes, and calculation of solution concentrations from titration results	2.7.1g (iii)	☐	☐
	Describe how to minimise measurement uncertainties in volumetric analysis and estimate overall uncertainty in a calculated result	2.7.1h	☐	☐

ResultsPlus
Build Better Answers

1   Sulfur dioxide is often used as a preservative in soft drinks. The amount used is critical – too little and harmful bacteria may grow in the drink; too much and the taste is unpleasant. The concentration of sulfur dioxide can be found by titration with iodine solution:

$$SO_2(aq) + I_2(aq) + 2H_2O(l) \rightarrow SO_4^{2-}(aq) + 2I^-(aq) + 4H^+(aq)$$

   a   Give the oxidation numbers of sulfur and iodine in the reactants and the products, and use them to show the type of reaction occurring. (4)

   b   25.0 cm³ of a drink reacted with 8.2 cm³ of 0.0050 mol dm⁻³ iodine solution. Find the concentration of sulfur dioxide in the drink in g dm⁻³. (4)

   c   Comment on the procedure used. (2)

✓ Examiner tip

   a   Sulfur increases from +4 (1) to +6 (1) so is oxidized (1), while iodine decreases from 0 to –1 (1) so is reduced and it is a redox reaction. (1)

     ■ **Basic answer:** Most students can correctly deduce the oxidation numbers. However, some students incorrectly write the oxidation numbers as S⁴⁺, S⁶⁺ etc. This would lose marks.

     ▲ **Excellent answer:** To get full marks for the explanation, students need to use the oxidation numbers to explain the redox nature of the reaction. Answers describing correct electron transfer would gain limited credit.

b Amount of iodine $= \frac{8.2}{1000}$ dm³ $\times 0.0050$ mol dm⁻³

$= 4.1 \times 10^{-5}$ mol   (1)

Amount of SO_2 equals amount of I_2, because 1 mol reacts with 1 mol in the balanced equation.

Amount of $SO_2 = 4.1 \times 10^{-5}$ mol   (1)

Concentration of $SO_2 = \frac{1000}{25.0} \times 4.1 \times 10^{-5}$ mol dm⁻³

$= 1.64 \times 10^{-3}$ mol dm⁻³   (1)

$= 1.64 \times 10^{-3}$ mol dm⁻³ $\times 64$ g mol⁻¹

$= 0.10$ g dm⁻³   (1)

Note that you will be penalised if you do not round your final answer to 2 significant figures (as in 8.2 and 0.0050), or give the units.

To two significant figures, $0.10496 = 0.10$

c The volumes are not similar/no evidence that the result is repeatable. (1) Should be repeated until there are two consistent or concordant results.   (1)

Practice exam questions

1 Write equations, with state symbols, for the following reactions:

  a sodium with oxygen   (2)

  b strontium with oxygen   (2)

  c calcium with water   (2)

  d magnesium with steam   (2)

  e sodium hydroxide with sulfuric acid   (2)

  f calcium hydroxide with hydrochloric acid   (2)

2 The following table shows the first four ionization energies of three elements in the same group of the Periodic Table – all the values are in kJ mol⁻¹.

Element	I.E.$_1$	I.E.$_2$	I.E.$_3$	I.E.$_4$
Q	738	1451	7733	10 541
T	590	1145	4912	6474
R	550	1064	1064	5500

  Q, **R** and **T** are **not** the actual symbols for the elements.

  a In which group of the Periodic Table would the elements belong?
    Give a reason for your answer.   (2)

  b Which element has the largest atomic number? Give a reason for your answer.   (2)

                (Adapted from Chemistry (Nuffield) CN1 Q4, June 2000)

3 In the redox reaction:

$$2I^-(aq) + Br_2(l) \rightarrow 2Br^-(aq) + I_2(s)$$

  which statement correctly describes what occurs during the reaction?   (1)

  A The I^- ion is oxidized, and its oxidation number decreases.

  B The I^- ion is oxidized, and its oxidation number increases.

  C The I^- ion is reduced, and its oxidation number increases.

  D The I^- ion is reduced, and its oxidation number decreases.

4 In the titration of hydrochloric acid with sodium hydroxide the following data were collected:

$$\text{volume of HCl used} = 14.4\text{ cm}^3$$
$$\text{volume of NaOH used} = 22.4\text{ cm}^3$$
$$\text{concentration of NaOH} = 0.20\text{ mol dm}^{-3}$$

  What is the concentration of the acid solution?   (1)

  A 0.64 mol dm⁻³   **B** 0.13 mol dm⁻³

  C 0.31 mol dm⁻³   **D** 1.6 mol dm⁻³

Group 7 and reactions of the halides

You should know the appearances of each halogen, and their solutions in water and in a hydrocarbon solvent:

- Chlorine is a pale yellow-green gas, which dissolves in water to give a pale yellow-green solution, and dissolves in a hydrocarbon solvent to give a pale yellow-green solution.
- Bromine is a red-brown liquid that is very volatile giving a red-brown gas. It is partially soluble in water and very soluble in a hydrocarbon solvent to give a red-brown solution in both situations.
- Iodine is a grey-black solid, which sublimes on heating to give a purple gas. It is slightly soluble in water giving a pale yellow solution. It is very soluble in a hydrocarbon solvent, giving a pink or red solution. It is soluble in potassium iodede solution to give yellow to brown solution depending on concentration.

The physical states of the halogens at room temperature reflect the strength of the intermolecular forces. As the number of electrons increase, the London forces increase in strength. Iodine, with 106 electrons, is a solid while chlorine, with 34 electrons, is a gas.

The halogens are more soluble in a hydrocarbon solvent than in water, because there are strong London forces between the solvent and the halogens. The halogens cannot form hydrogen bonds with water so are only slightly soluble.

Oxidation reactions

The halogens are **oxidizing agents** – with their strength decreasing down the group. It is important to remember that a halogen can oxidize other halide ions if the halide is below it in the group:

- chlorine water oxidizes bromide ions to bromine, and iodide ions to iodine
- bromine water oxidizes iodide ions to iodine
- iodine is not a strong enough oxidant to oxidize chloride ions or bromide ions.

You should be able to write ionic equations for these reactions like:

$$Br_2(aq) + 2I^-(aq) \rightarrow 2Br^-(aq) + I_2(aq)$$

The halogens can oxidize metals and non-metals. For example, with iron or phosphorus:

$$3Cl_2(g) + 2Fe(s) \rightarrow 2FeCl_3(s) \qquad 3Cl_2(g) + 2P(s) \rightarrow 2PCl_3(l)$$

In excess chlorine, phosphorus(V) chloride is formed:

$$5Cl_2(g) + 2P(s) \rightarrow 2PCl_5(s)$$

Chlorine also oxidizes iron(II) to iron(III), causing a pale green solution to turn brown:

$$Fe^{2+}(aq) + \tfrac{1}{2}Cl_2(aq) \rightarrow Fe^{3+}(aq) + Cl^-(aq)$$

Disproportionation of the halogens

Chlorine **disproportionates** in cold dilute alkali, forming a mixture of chloride (Cl^-) and chlorate (ClO^-) ions:

$$Cl_2(aq) + 2OH^-(aq) \rightarrow Cl^-(aq) + ClO^-(aq) + H_2O(l)$$

In cold dilute sodium hydroxide:

$$Cl_2(aq) + 2NaOH(aq) \rightarrow NaCl(aq) + NaOCl(aq) + H_2O(l)$$
$$ 2(0) -1 +1$$

Notice how chlorine both increases (0 to +1) and decreases (0 to −1) in oxidation number.

With hot concentrated alkali, a further decomposition takes places to form halate(V) ions. For example, with hot concentrated potassium hydroxide, solid iodine forms a mixture of potassium iodate(V) and potassium iodide:

$$3I_2(aq) + 6KOH(aq) \rightarrow KIO_3(aq) + 5KI(aq) + 3H_2O(l)$$
$$ 6(0) +5 5(-1)$$

Notice how iodine increases its oxidation number by five units from 0 in I_2 to iodate(V), and decreases by one unit from 0 in I_2 to iodide – so the reduction has to happen 5 times to balance the electrons.

Test for halides: the silver halides

Different halide solutions form different coloured precipitates with silver nitrate solution. Adding aqueous ammonia solution can confirm the identification, because different halide precipitates have different solubilities in aqueous ammonia.

- Chloride ions give a *white* precipitate of silver chloride, which darkens in sunlight. Also the precipitate dissolves to form a more stable compound in *dilute* ammonia solution.
- Bromide ions give a *cream* precipitate of silver bromide, which darkens in sunlight. Also the precipitate dissolves to form a more stable compound in *concentrated* ammonia solution.
- Iodide ions give a *yellow* precipitate of silver iodide, which does not darken in sunlight and does not dissolve in concentrated ammonia solution.

You should be able to write ionic equations for the formation of these precipitates. For example:

$$Ag^+(aq) + I^-(aq) \rightarrow AgI(s)$$

Reaction of the potassium halides with concentrated sulfuric acid

All the halogens react with concentrated sulfuric acid to give a hydrogen halide. The reactions show that there is a **trend** in the strength of the halide ions as **reducing agents**:

$$I^- > Br^- > Cl^-.$$

- Potassium chloride reacts to give hydrogen chloride, a gas that forms steamy fumes in moist air, and potassium hydrogensulfate. (Potassium sulfate only forms on prolonged heating at high temperature.)

$$H_2SO_4(l) + KCl(s) \rightarrow HCl(g) + KHSO_4(aq)$$

- Potassium bromide reacts initially in a similar way to give hydrogen bromide, but the hydrogen bromide formed is oxidized by the sulfuric acid to bromine, and sulfuric acid is reduced to sulfur dioxide (turns acidified potassium dichromate(VI) paper from yellow to green).
- With potassium iodide, the hydrogen iodide formed is immediately oxidized to iodine. The sulfuric acid is reduced to hydrogen sulfide (smells of 'rotten eggs' and turns lead ethanoate paper from white to black).

Reactions of the hydrogen halides

To make pure samples of hydrogen bromide and hydrogen iodide from their potassium salts, phosphoric(V) acid is used – it is not such a strong oxidizing agent as sulfuric acid.
- All three hydrogen halides fume in moist air.
- All three are extremely soluble in water, forming acidic solutions.
- All three react with ammonia (NH_3) to form a white smoke of the corresponding ammonium halide. For example, with hydrogen chloride:

$$NH_3(g) + HCl(g) \rightarrow NH_4Cl(s)$$

Fluorine and astatine

Given the trends in physical and chemical properties of chlorine, bromine and iodine, you should be able to predict the properties of fluorine and astatine and their compounds. Going down the group the halogens get less reactive, so fluorine reacts with just about everything, and astatine is less reactive than iodine.
- Fluorine should be a gas and astatine should be a solid – because an increase in the number of electrons increases the strength of London forces. Similarly, astatine will have the highest boiling temperature.
- Electronegativity decreases down the group, so astatine will have a low electronegativity value.
- Fluorine will be the most oxidizing.
- The compounds of fluorine and astatine with hydrogen should be soluble in water, forming acids.
- Bond enthalpies for carbon–halogen bonds decrease down the group. Fluorocarbon compounds should be the most stable, and astatocarbon compounds the least stable.

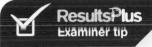

Quick Questions

1 Write ionic equations for the reactions of bromine water with:
   a a solution of potassium iodide
   b cold potassium hydroxide solution.
2 Write equations for the reaction of:
   a hydrogen iodide with ammonia
   b the thermal decomposition of hydrogen iodide.
   Describe what you would see during each reaction.

Iodine–thiosulfate titrations

Titration can be used to find the concentration of an iodine solution. It is titrated with sodium thiosulfate solution of known concentration. A typical question is to calculate the **percentage purity** of a sample of impure potassium iodate.

The reaction between iodine and thiosulfate ions

To learn the formula of the thiosulfate ion, remember that 'thio' means 'replace an oxygen with a sulfur in the formula' – so sulfate, SO_4^{2-}, becomes **thiosulfate, $S_2O_3^{2-}$**. The reaction between iodine and thiosulfate is a **redox reaction**. Iodine is reduced to iodide ions and thiosulfate is oxidized:

$$I_2(aq) + 2e^- \rightarrow 2I^-(aq) \qquad \text{reduction}$$
$$2S_2O_3^{2-}(aq) \rightarrow S_4O_6^{2-}(aq) + 2e^- \quad \text{oxidation}$$

Adding these gives the full ionic equation:

$$I_2(aq) + 2S_2O_3^{2-}(aq) \rightarrow 2I^-(aq) + S_4O_6^{2-}(aq)$$

(Notice the unusual oxidation number of sulfur in the tetrathionate ion, $+2.5$!)

This reaction can be used to work out the concentration of any oxidizing agent that will react with iodide ions to form iodine ... including potassium iodate(V).

A known volume of the oxidizing agent, potassium iodate(V), is reacted with iodide ions in *excess* potassium iodide solution to liberate iodine. The potassium iodide is in excess so that all of the iodate(V) reacts:

$$IO_3^-(aq) + 5I^-(aq) \rightarrow 3I_2(aq) + 3H_2O(l)$$

The iodine that is formed is then titrated with sodium thiosulfate. The colour change in the titration is yellow to colourless. The **accuracy** of the end-point can be enhanced by adding **starch** as an indicator just before the end-point. The blue-black colour disappears at the end-point, showing that all the iodine has *just* been reacted.

The preparation of potassium iodate(V)

You should know how to prepare potassium iodate(V):

- For example, calculate that the amount of iodine needed to react with 10 cm^3 of 4 mol dm^{-3} potassium hydroxide solution is 5.12 g.
- Iodine is added until the colour is just visible.
- Further drops of potassium hydroxide solution are added until the brown colour turns very pale yellow.
- Crystals are formed by precipitation from a hot solution. On cooling, potassium iodate(V) crystallizes before potassium iodide because it is much less soluble in cold water. However, the sample is impure.
- The potassium iodate(V) is collected by suction filtration and dried between filter papers.

Finding the purity of a sample of potassium iodate(V)

You need to remember the details of how to make up a **standard solution** of a known mass (about 0.5 g) of impure potassium iodate(V):

- Weigh a weighing bottle with lid.
- Weigh the weighing bottle containing about 0.5 g of your sample of potassium iodate(V).
- Tip the contents of your weighing bottle through a funnel into a **volumetric flask**.
- Rinse your weighing bottle with water, and pour the rinsings through the funnel into the flask.
- Make up the flask to the 100 cm³ graduation mark.
- Shake the flask thoroughly.
- Pipette 10 cm³ of this solution into a 100 cm³ conical flask.
- Titrate the iodine formed with 0.010 mol dm⁻³ sodium thiosulfate solution.
- Calculate the concentration of the iodine solution, and hence the percentage purity of your sample of potassium iodate(V).

Titration calculation

The steps in this calculation are similar to those in an acid–base titration.

Step 1.   Calculate the number of moles of sodium thiosulfate used in the titration, from the known volume and concentration.

Step 2:   Calculate the number of moles of iodine that reacted with this.

$$I_2(aq) + 2S_2O_3^{2-}(aq) \rightarrow 2I^-(aq) + S_4O_6^{2-}(aq)$$

The equation for the reaction shows that 1 mole of iodine reacts with 2 moles of thiosulfate, so you need to divide the number of moles of thiosulfate by 2.

Step 3:   The equation for the reaction between iodate(V) and iodide ions is:

$$IO_3^-(aq) + 5I^-(aq) \rightarrow 3I_2(aq) + 3H_2O(l)$$

So, to find the number of moles of iodate(V) in 10 cm³, you need to divide the number of moles of iodine by 3.

Step 4:   To find the number of moles of potassium iodate in the solid sample, it is the amount in the 100 cm³ volumetric flask – so multiply by 10 because 10 cm³ of the 100 cm³ solution was taken in the pipette.

Step 5:   To find the mass, multiply by the molar mass of potassium iodate(V).

Step 6:   To calculate the percentage purity, divide by the mass of sample and multiply by 100.

> **ResultsPlus**
> **Examiner tip**
>
> As with other titration calculations, you will probably receive some guidance in this type of calculation in an AS examination. The question will usually be broken down into shorter steps.

?) Quick Questions

1 Write the equation for the formation of potassium iodate(V) from iodine and potassium hydroxide solution. Describe how you would carry out this reaction in the laboratory.

2 25 cm³ of potassium iodate(V) solution was added to excess potassium iodide solution. The iodine liberated required 30 cm³ of 0.50 mol dm⁻³ sodium thiosulfate. Use the equations given on this page to calculate the concentration of the potassium iodate(V) solution.
   A 0.1 mol dm⁻³   B 0.2 mol dm⁻³
   C 0.4 mol dm⁻³   D 0.5 mol dm⁻³

Reaction rates and catalysts

Reaction kinetics forms the study of rates of reaction and the factors which change them. At AS, a qualitative overview is all that is needed. At A2 a detailed quantitative understanding is needed.

Factors influencing reaction rate

The factors which influence the rate of a reaction are:
- **Concentration** – increasing the concentration of solutions increases the **frequency of collisions** between particles, which increases the reaction rate.
- **Pressure** – increasing the pressure in a gas reaction has the same effect as increasing concentration, for the same reason.
- **Temperature** – increasing the temperature increases the reaction rate. This is partly because it increases the collision frequency, but there is an important additional reason described later.
- **Surface area** – increasing the surface area of a solid increases the reaction rate. For some gas reactions, the provision of a surface helps the gas molecules to come together and weakens their bonds, increasing the reaction rate.
- **Catalysts** – increase the reaction rate.

For a reaction to occur three conditions are needed:
- a collision between the reacting particles
- the reacting particles must have sufficient energy to break their existing bonds, so that new bonds can form
- they must collide in the correct orientation.

The energy needed to break the existing bonds is called the **activation energy** for a reaction.

The effect of temperature – distribution of molecular energies

The effect of temperature on a reaction mixture is to increase the energy of the particles. At a given temperature, the particles do not have one particular energy – but a range of energies. When the temperature increases, the distribution of **molecular energies** changes. This distribution is known as the **Maxwell–Boltzmann distribution** and is shown for two different temperatures in the diagram below.

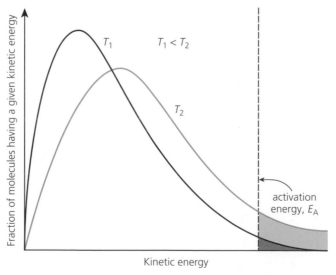

Kinetic energy distributions of the particles for a reaction mixture at two different temperatures. The shaded areas are proportional to the total fraction of particles that have the minimum activation energy. You can see that at the higher temperature, T_2, more particles have enough energy to react

Note that the x-axis measures energy; the y-axis measures the fraction of particles with a particular energy. Notice that:
- *no* molecules have zero energy
- each curve rises to a peak, and then falls away to approach the x-axis without ever meeting it
- at higher temperatures the peak of the graph is lower and moves to the right
- the area under the curve is the same at both temperatures because there are the same number of particles.

You should be able to sketch this graph and to explain its significance. At higher temperatures, there are more molecules with higher energy – and in particular *there are many more molecules with energy higher than the activation energy*. This is the main reason why increasing the temperature of a reaction increases the reaction rate.

The role of catalysts

At AS we have a new definition for a catalyst:
- a catalyst provides an **alternative route** for a reaction, which has a **lower activation energy**.

It is important that you know this new definition – both parts are needed!

The effect of a catalyst can be shown on a **reaction profile diagram** – particularly when the diagram shows the profiles for both the uncatalysed reaction and the catalysed reaction.

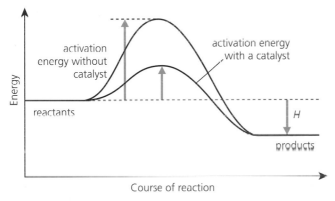

A catalyst lowers the activation energy of a reaction by forming an activated complex, a low-energy intermediate stage

Notice how this reaction profile diagram shows how the activation energy is less for the catalysed reaction than for the uncatalysed reaction. This particular profile is for an exothermic reaction – which is why the products are at lower energy than the reactants.

ResultsPlus
Watch out!

You will often be asked to sketch labelled reaction profiles. Take care to show the relative energy levels of the reactants and products the right way round. Students lose marks by drawing the activation energy hump as straight lines, or failing to label the activation energy of both the catalysed and uncatalysed reactions.

Quick Questions

1. **a** Explain why increasing the pressure increases the rate of a reaction between gases.
   b Suggest two disadvantages of working at high pressures.
2. As temperature increases, describe how the Maxwell–Boltzmann distribution changes. Use this to explain why increasing the temperature has such a dramatic effect on reaction rate.

Chemical equilibria

Chemical equilibria can only be established in closed systems. They have three key features – they are:

- reactions that do not go to **completion** – at the 'end' of the reaction there are both products and reactants present
- reactions that are **reversible** – if you start with the products, some reactants form, and vice versa
- **dynamic** – when the reaction appears to have finished, the chemicals continue to react, but the **rate** of the forward reaction is equal to the rate of the reverse reaction. This is called a **dynamic equilibrium**.

The effect of changes in conditions on equilibria

These are determined by applying **Le Chatelier's principle**. This states that:

- when a change is imposed on a chemical equilibrium, the reaction responds in such a way as to oppose the change. As a result, the **position of equilibrium** changes.

Worked Example

This can be demonstrated in many ways – for example, the red-brown gas NO_2 exists in equilibrium with pale yellow N_2O_4:

$$N_2O_4(g) \rightleftharpoons 2NO_2(g)$$
$$\text{yellow} \qquad \text{brown}$$

If the position of equilibrium shifts to the *left*, the mixture pales – you get more of the *left-hand* reactant N_2O_4.

If the position of equilibrium shifts to the *right*, the mixture darkens – you get more of the *right-hand* product NO_2.

Change in temperature

The sign of ΔH, the **enthalpy change** for the reaction, is needed.

- If the temperature of an 'exothermic forward' equilibrium reaction is increased, the reaction will respond by going in the direction that lowers the temperature – and the reverse reaction will be favoured.
- If the temperature of an 'endothermic forward' equilibrium reaction is increased, the reaction will respond by going in the direction that lowers the temperature – and the forward reaction will be favoured.

The reversible reaction between N_2O_4 and NO_2 is endothermic in the forward direction:

$$N_2O_4(g) \rightleftharpoons 2NO_2(g) \qquad \Delta H = +58.1 \, \text{kJ mol}^{-1}$$

- When a closed tube containing the equilibrium mixture is put into hot water, the equilibrium is shifted to favour the production of NO_2 – the colour darkens.
- When the temperature is decreased, the equilibrium is shifted to favour the production of N_2O_4 – the colour pales.

Change in pressure

The pressure of a gas depends on the number of molecules in its container. If the pressure of a gaseous equilibrium reaction is increased, the reaction will go in the direction that will reduce the pressure – the direction that reduces the number of gaseous molecules. The reverse argument applies to decreasing the pressure. In the reaction:

$$N_2O_4\,(g) \rightleftharpoons 2NO_2\,(g)$$

- an increase in pressure favours the production of N_2O_4 because that is the side with the smaller number of molecules – the colour pales.
- a decrease in pressure favours the production of NO_2, and an increase in the number of molecules – the colour darkens.

Notice this means that pressure affects only equilibrium reactions in which there is a change in the number of gaseous molecules.

Change in concentration

- Lowering the concentration of a reactant will make the reaction reverse, in attempting to make more of that reactant.
- Increasing the concentration of a reactant will make the reaction go forwards, in attempting to reduce the concentration of that reactant.

For example:

$$PCl_5(g) \rightleftharpoons PCl_3(g) + Cl_2(g)$$

If the concentration of $Cl_2(g)$ is increased, the equilibrium shifts to the *left* to try to reduce the concentration of $Cl_2(g)$, so more phosphorus(V) chloride is formed.

ResultsPlus
Examiner tip

Most AS examination equilibrium questions are about conditions of temperature and pressure. The most common errors are failure to use and apply the terms endothermic and exothermic correctly with temperature changes, and the omission of the word gaseous in explaining the effect of pressure favouring the reaction that produces the most gaseous molecules.

Worked Example

State the theoretical conditions of temperature and pressure that favour the production of sulfur trioxide from sulfur dioxide and oxygen:

$$2SO_2(g) + O_2(g) \rightleftharpoons 2SO_3(g) \qquad \Delta H = -197\,kJ\,mol^{-1}$$

To favour production of sulfur trioxide, a high pressure would be needed because that will favour the forward reaction, which lowers the number of gaseous molecules. (In practice, the reaction is carried out close to atmospheric pressure because producing high pressure is very costly.)

Also, lowering the temperature would make the exothermic reaction go in the forward direction to give out heat and try to increase the temperature. (However, this would decrease the rate of the reaction.)

Quick Questions

1 You may have seen the formation of the yellow solid iodine trichloride from the brown liquid iodine monochloride and chlorine gas. When you tip a U-tube containing yellow iodine monochloride, the residual chlorine gas pours out and the solid turns to a brown liquid. Explain why.

2 State the preferred conditions of temperature and pressure to favour the production of ammonia from nitrogen and hydrogen:

$$N_2(g) + 3H_2(g) \rightleftharpoons 2NH_3(g) \qquad \Delta H = -92\,kJ\,mol^{-1}$$

Topic 3: Inorganic Chemistry 2 – Group 7, equilibria and reaction rates checklist

By the end of this topic you should be able to:

Revision spread	Checkpoints	Specification section	Revised	Practice exam questions
Group 7 and reactions of the halides	Recall the appearance and states of the Group 7 elements and their solutions in water and hydrocarbon solvents	2.7.2a	☐	☐
	Describe the reactions of halogens: (i) oxidation of metals, their ions and non-metals (ii) disproportionation with cold and hot alkali solutions	2.7.2b	☐	☐
	Predict the physical properties and chemistry of fluorine and astatine	2.7.2d	☐	☐
Iodine/ thiosulfate titrations	Know about iodine/thiosulfate titrations, including the indicator with colour changes and the calculation of results. Describe the following reactions: (i) potassium halides with concentrated sulfuric acid, halogens and silver nitrate solution (ii) silver halides with sunlight and their solubilities in ammonia solution (iii) hydrogen halides with water and ammonia	2.7.2c	☐	☐
Reaction rates and catalysts	State the five factors that influence reaction rate	2.8a	☐	☐
	Use collision theory to explain changes in reaction rates	2.8b	☐	☐
	Sketch the Maxwell–Boltzmann distribution of molecular energies and relate it to the effect of concentrations and temperature on reaction rate	2.8c	☐	☐
	Explain what is meant by activation energy, and its relation to the effect of temperature changes on reaction rate	2.8d	☐	☐
	Describe how catalysts work in general terms and draw reaction profiles for catalysed and uncatalysed reactions	2.8e	☐	☐
	Describe experiments that demonstrate the factors influencing reaction rates	2.8f	☐	☐
Chemical equilibria	Appreciate that equilibrium is a dynamic state	2.9a	☐	☐
	Deduce the qualitative effects of changes in temperature, pressure and concentration on the position of equilibrium reactions	2.9b	☐	☐
	Interpret the results of experiments which show the changes in 2.9b	2.9c	☐	☐

ResultsPlus
Build Better Answers

1 a Concentrated sulfuric acid is not suitable for preparing pure samples of hydrogen halides – HBr and HI – from their corresponding potassium halides. Explain why concentrated sulfuric acid is not suitable. (1)

b A test tube containing hydrogen chloride gas is inverted in water. Describe and explain what you would see. (2)

c i What would you see when ammonia reacts with hydrogen bromide? (1)

  ii What is the formula of the product of the reaction between ammonia and hydrogen bromide? (1)

d A hot wire is plunged into a test tube containing each of the hydrogen halides.

  i The bond energies of the hydrogen halides are given in the table: Suggest and explain which of the hydrogen halides will decompose. (2)

  ii Describe the appearance of the contents of the test tube when decomposition takes place. (1)

  iii Write a balanced equation for the reaction for the decomposition in **(i)**. (1)

Hydrogen halide	Bond energy/kJ mol^{-1}
HF	568
HCl	432
HBr	366
HI	298

Examiner tip

a Sulfuric acid oxidizes the products hydrogen bromide and hydrogen iodide. (1)
Tell the examiner all you know – not just 'sulfuric acid oxidizes' but which chemical it oxidizes as well.

b The water rushes up the test tube (1) because hydrogen chloride is extremely soluble in water. (1)

c i White smoke. (1) Note, the question asks what you would *see*, so you would not get credit for stating the name of the product as ammonium bromide.

ii NH_4Br (1)

Make sure you are clear on the formulae for ammonia and ammonium; students often confuse NH_3 and NH_4^+

d i Hydrogen iodide/HI (1) because it has the weakest bond (breaks most easily). (1)

ii A purple gas condenses to a black solid. (1) Always state the appearance before and after or you will lose the mark.

iii $2HI(g) \rightarrow H_2(g) + I_2(g/s)$ (1)

(Adapted from Chemistry (Nuffield) CN2 Q1, Jan 96)

Practice exam questions

1 Solid ammonium chloride exists in equilibrium with ammonia and hydrogen chloride gases at 25°C:

$$NH_4Cl(s) \rightleftharpoons NH_4(g) + HCl(g) \quad \Delta H = +92.42\ kJ\ mol^{-1}$$

What change will shift the equilibrium to the right? (1)

A decreasing the temperature to 15°C

B increasing the temperature to 35°C

C removal of $NH_4Cl(s)$

D addition of a catalyst

2 Which element cannot be displaced from a solution of its potassium salt by the action of chlorine? (1)

A fluorine **B** bromine **C** iodine **D** astatine

3 a An aqueous solution of silver nitrate is added to an aqueous solution of potassium iodide in a test tube.

i State what you would observe. (2)

ii Write a balanced ionic equation, including state symbols, for the reaction that occurs. (2)

b The reaction in **(a)** is repeated using potassium bromide instead of potassium iodide. The reaction mixture is allowed to stand in sunlight for a few minutes.

i State what you would observe. (1)

ii Suggest a useful application of this reaction. (1)

iii Concentrated ammonia solution is added to the reaction mixture. State and explain what you would observe. (2)

c A few drops of concentrated sulfuric acid are added to some potassium bromide in a test tube. At first a gas is given off which fumes at the mouth of the test tube, and gives dense white smoke with ammonia.

i Name the gas given off. (1)

ii Write a balanced equation for the reaction between the gas and ammonia. (1)

iii After a short time, reddish-brown fumes are observed in the test tube. Name this gas. (1)

iv During this time a piece of filter paper soaked in acidified sodium dichromate(VI) is held in the end of the test tube. The colour changes from yellow to green. Name the gas that causes this change. (1)

v Concentrated sulfuric acid has reacted in two ways in these reactions. Classify its behaviour in reactions **(c)(i)** and **(c)(iii)**. (2)

(From Chemistry (Nuffield) CN2Q1, June 95)

Alcohols

A simple alcohol is an alkane in which one hydrogen is replaced by the **OH functional group**.

Naming alcohols

Name as for alkanes (see page 50) except:
- the names end in -ol instead of –ane, so ethane becomes ethanol
- you need to give the position of the OH group (the number of the carbon atom it is attached to) before -ol in the usual way.

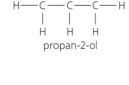
propan-2-ol

Worked Example

What is the name of $CH_3CH(OH)\Delta$?

Step 1: There are three carbon atoms in a continuous chain, so the stem is prop-

Step 2: The OH group is attached to the second carbon atom, so put -2 in front of -ol

Step 3: The name is propan-2-ol.

You need to remember the meaning of the terms **primary**, **secondary** and **tertiary**. The key is the number of carbons attached to the carbon that is directly attached to the OH group:
- one carbon atom attached makes it primary – as in propan-1-ol, $CH_3CH_2CH_2OH$
- two makes it secondary – as in propan-2-ol, $CH_3CHOHCH_3$
- three makes it tertiary – as in 2-methylpropan-2-ol, $(CH_3)_3COH$.

If there is more than one alcohol group then use 'di' or 'tri', as appropriate – as in propane-1,2,3-triol (also known as glycerol).

Reactions of alcohols

For each of the reactions, you should be able to give:
- the names of the reactants and the conditions for the reaction
- any obvious changes you would see during the reaction
- the names of the products
- the formulae of products – displayed, structural and skeletal
- one test for each product.

You should be able to do this for any alcohol containing up to four carbon atoms.

Reaction with sodium

Alcohols react with solid sodium. The sodium disappears, bubbles of gas form, and a white solid product forms. For example, with propan-1-ol:

$$CH_3CH_2CH_2OH(l) + Na(s) \rightarrow CH_3CH_2CH_2O^-Na^+(s) + \tfrac{1}{2}H_2(g)$$

The white, solid organic product is sodium propoxide, which is ionic. The gas is hydrogen, but you must not give 'hydrogen forms' as an observation because you can't distinguish hydrogen from any other colourless gas by sight.

Formation of halogenoalkanes

Halogenoalkanes can be prepared from primary, secondary and tertiary alcohols by incorporation of a halogen atom. All alcohols react with phosphorus(V) chloride (PCl_5) or with concentrated hydrochloric acid to form chloroalkanes. For example, with butan-1-ol, 1-chlorobutane is formed:

$$PCl_5(s) + CH_3CH_2CH_2CH_2OH(l) \rightarrow CH_3CH_2CH_2CH_2Cl(l) + HCl(g) + POCl_3(l)$$
$$HCl(aq) + CH_3CH_2CH_2CH_2OH(l) \rightarrow CH_3CH_2CH_2CH_2Cl(l) + H_2O(l)$$

Notice that both are **substitution reactions**, because a chlorine atom Cl has replaced an OH group.

The first of these reactions is used as the test for the OH group in an organic compound.

ResultsPlus
Examiner tip

If an organic liquid reacts with phosphorus(V) chloride to produce steamy fumes of hydrogen chloride, which turn blue litmus red, the OH group is present.

Combustion

The simplest oxidation reaction of alcohols is **combustion** to release energy. Complete combustion results in the formation of carbon dioxide and water vapour only. For example:

$$C_2H_5OH(l) + 3O_2(g) \rightarrow 2CO_2(g) + 3H_2O(g)$$

Oxidation reactions

The oxidizing agent is acidified sodium or potassium dichromate(VI), which changes from orange to a dark green as the dichromate(VI) is reduced to chromium(III). The acid used is sulfuric acid.

- If dilute sulfuric acid is used and the product is **distilled** immediately, then primary alcohols form aldehydes. For example, butan-1-ol forms butanal:

$$3CH_3CH_2CH_2CH_2OH(l) + Cr_2O_7^{2-}(aq) + 8H^+(aq) \rightarrow 3CH_3CH_2CH_2CHO(l) + 2Cr^{3+}(aq) + 7H_2O(l)$$

Aldehydes form a red precipitate when boiled with Benedict's or Fehling's solution.

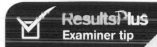

Results Plus
Examiner tip

You do not need to be able to balance these redox equations at AS.

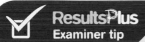
Results Plus
Examiner tip

In the Unit 3 practical assessment, you may have to prepare an organic liquid. This could be an aldehyde, carboxylic acid, ketone or halogenoalkane. To prepare an aldehyde by oxidation of an alcohol you must distil it off as it forms, or it will oxidize further to a carboxylic acid.

In the written exam, you must be able to draw the apparatus needed for the processes of distillation and reflux. Remember to draw the jacket of your condenser and to mark the water direction coming in at the bottom of the condenser. You will lose marks if you show either apparatus stoppered or sealed.

- If concentrated sulfuric acid is used and the mixture is heated under reflux before being distilled, further oxidation takes place to form a carboxylic acid. For example, butan-1-ol forms butanoic acid:

$$3CH_3CH_2CH_2CH_2OH(l) + 2Cr_2O_7^{2-}(aq) + 16H^+(aq) \rightarrow 3CH_3CH_2CH_2COOH(l) + 4Cr^{3+}(aq) + 11H_2O(l)$$

A useful test for carboxylic acids is that they will neutralize a large volume of sodium carbonate solution.

Note that reflux involves continual boiling and condensing of the reactant to ensure that the reaction takes place but without the contents of the flask boiling dry. The condenser prevents loss of the solvent by evaporation.

- When secondary alcohols are oxidized with acidified sodium dichromate(VI), ketones are formed, which cannot be oxidized further. For example, butan-2-ol forms butanone, $CH_3CH_2COCH_3$.

Ketones do not react with Benedict's solution, but they do give a yellow/orange precipitate with Brady's reagent.

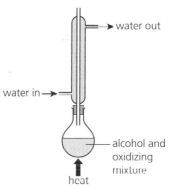

water out

water in

alcohol and oxidizing mixture

heat

Reflux apparatus (vertical condenser) for preparing a carboxylic acid from a primary alcohol

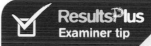
Results Plus
Examiner tip

The reason you prepare a carboxylic acid by heating under reflux is so that you can heat the alcohol for a prolonged time without losing volatile reactants or products.

? Quick Questions

1 Give the names for the two primary alcohols with the molecular formula $C_4H_{10}O$.
2 **a** Write the equation, including state symbols, for the reaction of ethanol with sodium.
   b Name the organic product.
3 **a** Give the names and structural formulae for the two possible products of the reaction of methanol with sodium dichromate(VI).
   b For each product, give a test that would confirm the presence of the functional group.

Mass spectra and infrared absorption

Interpreting mass spectra of organic compounds

The principles of operation of a **mass spectrometer** were described in Unit 1 (pages 28–29). It is important to recall that:

- the peak furthest to the right (highest m/z value) in the **mass spectrum** corresponds to the ion with the largest mass – i.e. the whole molecule ionized
- z is usually taken to be 1 (charge $+1$ corresponding to the loss of one electron), so the highest m/z particle corresponds to the relative molecular mass of the compound
- all the ions produced in a mass spectrometer are *positive*
- because the molecular ion is unstable, it breaks into smaller **fragment ions** of different masses
- these fragments produce other peaks, which can often help to identify the structure of the compound or to distinguish between compounds – the ion peaks indicate likely **functional groups**.

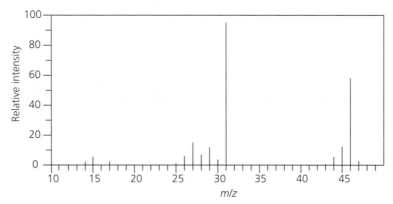

The mass spectrum of ethanol

In the mass spectrum of ethanol:

- the molecular ion $C_2H_5OH^+$ is readily identified as having mass 46
- the peak at 45 must be due to $C_2H_5O^+$ (-1 mass, so loss of H)
- the peak at 15 is due to CH_3^+
- the peak at 29 is due to $C_2H_5^+$
- the peak at 31 is due to CH_2OH^+.

Worked Example

How would mass spectra distinguish between the **isomers** propan-1-ol ($CH_3CH_2CH_2OH$) and propan-2-ol ($CH_3CH(OH)CH_3$)?

- -

- Both would have peaks at m/z 60, corresponding to the parent ion $C_3H_8O^+$
- Both would have a peak at m/z 48 due to the $CH_3CH_2CH_2^+$ and $CH_3CHCH_3^+$ ions
- Only propan-1-ol would have a peak at m/z 29 due to $CH_3CH_2^+$.

Infrared spectroscopy

Molecules absorb infrared radiation (IR) by vibrating with increased kinetic energy. All chemical bonds have a **vibrational frequency**, which depends on the bond strength. Different bonds absorb radiation at specific frequencies, or wavelengths, in the IR region of the electromagnetic spectrum – for each, a '**peak**' (actually a trough!) occurs in the **absorption spectrum**. Note that transmittance increases upwards in an IR spectrum printout, so 'troughs' correspond to *high* levels of absorbance. This is why the troughs are called peaks.

The unit of measurement of absorbance is the number of waves per centimetre (cm^{-1}) or **wavenumber** – this is just the reciprocal of the wavelength.

Interpreting infrared spectra

The main use of **IR spectroscopy** is in the determination of molecular structure. Here it is useful to know where to start, and where to look for **bonds** in key organic **functional groups**.

- Start by looking around wavenumber 3000 cm^{-1}.
- **C—H bonds** in an alkene occur as a *single* absorption just above 3000 cm^{-1}; for an alkane just below 3000 cm^{-1}.
- **O—H bonds** in alcohols absorb at about 3600 cm^{-1}. This is nearly always a broad absorption (over 100–200 cm^{-1}) due to hydrogen bonding.
- In carboxylic acids, the peak around 3600 cm^{-1} due to the **O—H bond** is even *broader* – it may stretch back to 2500 cm^{-1} quite obliterating anything else!
- The carbonyl bond (**C=O**) peak occurs around 1700 cm^{-1}. It is highest in aldehydes at 1720–1740 cm^{-1}. Ketone carbonyl bonds come just below aldehydes at 1680–1700 cm^{-1}.
- **N—H bonds** in amines absorb at about 3500–3300 cm^{-1}, but the peak is not as strong or as broad as for O—H bonds.
- Do not worry about anything else! The right-hand side of the spectrum is the 'fingerprint' region, which can be used to identify a specific compound, but you will not usually be asked to do this at AS.

In the diagram below there is a strong, broad peak between 3230 and 3500 corresponding to the OH group in an alcohol.

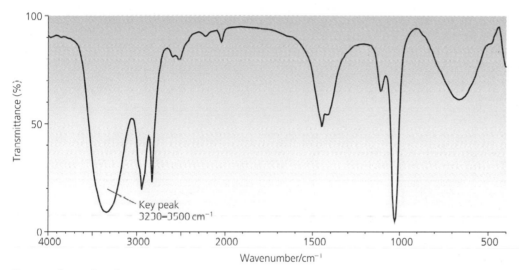

IR spectra for methanol

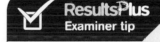

For an exam, you do not need to memorise IR absorption data – you may be given wavenumber data for some familiar bonds and the functional groups they correspond to. You must be able to use any data provided to interpret spectra. You should try both identifying compounds from their IR spectra and predicting the spectra of known compounds.

Absorption of IR and greenhouse gases

Only molecules that change their polarity as they vibrate (due to the movement of the dipoles in the polar bonds) absorb infrared radiation. So water, carbon dioxide, methane and carbon monoxide absorb IR, but oxygen and nitrogen do not. Only molecules that can absorb IR are greenhouse gases.

Quick Questions

1. Explain how IR spectroscopy can be used to distinguish between ethanol, ethanal and ethanoic acid.
2. Explain how you would distinguish 1-bromobutane from 2-bromobutane using their mass spectra.
3. Suggest a likely ion fragment for ethanol, and at what *m/z* value this would occur.
4. The mass spectrum for propanal includes peaks at *m/z* values 29, 57 and 58.
   a. Explain why the mass spectrum shows peaks at these values.
   b. Suggest how the mass spectrum for propanone would differ.
5. Which of the following are greenhouse gases?
   chlorine, chloromethane, nitrogen dioxide, argon, ammonia

Topic 4: Organic chemistry 1 – alcohols and spectroscopy checklist

By the end of this topic you should be able to:

Revision spread	Checkpoints	Specification section	Revised	Practice exam questions
Alcohols	Recognise the alcohol functional group	2.10.1a	☐	☐
	Name alcohols, and draw their displayed, structural and skeletal formulae. Classify alcohols as primary, secondary or tertiary	2.10.1b	☐	☐
	Describe the following reactions of alcohols: combustion; reaction with sodium; substitution with halogen to form halogenoalkanes, including the test for alcohols; oxidation of primary alcohols to produce aldehydes and carboxylic acids, and secondary alcohols to produce ketones	2.10.1c	☐	☐
	Prepare an organic liquid using reflux and distillation, and purify an organic liquid	2.10.1d	☐	☐
Mass spectra and infrared absorption	Interpret fragment ion peaks in mass spectra	2.12a	☐	☐
	Use infrared spectra to deduce functional groups present, or predict infrared spectra from a knowledge of functional groups present limited to C—H absorptions in alkanes, alkenes and aldehydes; O—H absorptions in alcohols and carboxylic acids; N—H absorptions in amines; C=O absorptions in aldehydes and ketones	2.12b	☐	☐
	Know how to work out and explain which molecules absorb infrared radiation	2.12c	☐	☐
	Know how to work out which gases are greenhouse gases	2.12d	☐	☐

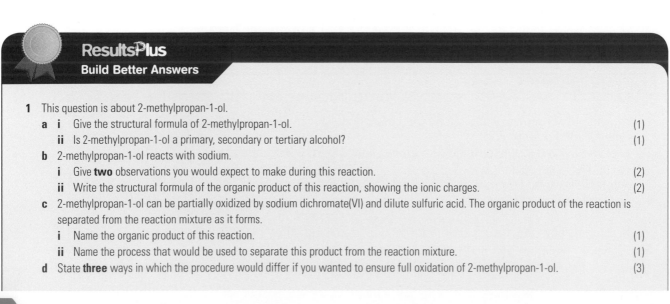

ResultsPlus
Build Better Answers

1 This question is about 2-methylpropan-1-ol.
   a i Give the structural formula of 2-methylpropan-1-ol. (1)
      ii Is 2-methylpropan-1-ol a primary, secondary or tertiary alcohol? (1)
   b 2-methylpropan-1-ol reacts with sodium.
      i Give **two** observations you would expect to make during this reaction. (2)
      ii Write the structural formula of the organic product of this reaction, showing the ionic charges. (2)
   c 2-methylpropan-1-ol can be partially oxidized by sodium dichromate(VI) and dilute sulfuric acid. The organic product of the reaction is separated from the reaction mixture as it forms.
      i Name the organic product of this reaction. (1)
      ii Name the process that would be used to separate this product from the reaction mixture. (1)
   d State **three** ways in which the procedure would differ if you wanted to ensure full oxidation of 2-methylpropan-1-ol. (3)

☑ **Examiner tip**

a i $CH_3CH(CH_3)CH_2OH$ (1)

ii Primary   (1)

Many students have difficulty classifying alcohols. Remember that the classification depends on the number of carbon atoms attached to the *carbon attached* to the alcohol group.

b i Any two from sodium disappears/dissolves, bubbles of gas form, a white solid forms   (2)

You may know that hydrogen forms, but you cannot *see* hydrogen gas – be careful to answer the question. Observations are what you can *see*.

ii $CH_3CH(CH_3)CH_2O^-Na^+$ formula (1)   charges (1)

c i 2-methylpropanal   (1)

ii Distillation   (1)

d Use concentrated sulfuric acid; (1) Use twice as much/excess sodium dichromate(VI); (1) Heat under reflux followed by distillation. (1)

■ **Basic answer:** Most students know that these compounds have the same molecular mass so will have the same largest *m/z* value/ parent ion. But the question asks you to *distinguish* the two spectra, so no marks are given for stating the similarities.

▲ **Excellent answer:** For full marks you need to predict the likely molecular ion fragments and work out their masses. This is the reverse of interpreting *m/z* peaks to deduce likely functional groups.

(Adapted from Chemistry (Nuffield) CN1 Q5, Jan 96)

Practice exam questions

1 Which of these is a tertiary alcohol?   (1)

  A 2-methylpropan-2-ol

  B propan-2-ol

  C pentan-3-ol

  D 2-methylpropan-1-ol

2 Bromine has two isotopes (^{79}Br and ^{81}Br) in an approximate 1 : 1 ratio. How many peaks will the parent ion contain in the mass spectrum of Br_2?   (1)

  A 1   **B** 2   **C** 3   **D** 4

3 This question is about propan-1-ol, some of whose reactions are shown below:

  Reaction 1   Reaction 2   Reaction 3

  $CH_3CH_2CH_2ONa \leftarrow CH_3CH_2CH_2OH \rightarrow CH_3CH_2CHO \rightarrow CH_3CH_2COOH$

  A   propan-1-ol   **B**   **C**

  a i Name the reactant which must be added to propan-1-ol to bring about Reaction 1.   (1)

  ii State the appearance of **A**, the product of Reaction 1.   (2)

  b The same reagents can be used in Reaction 2 and Reaction 3, but under different conditions.

    i What is the name for this type of reaction?   (1)

    ii Give the names of the two reagents suitable for both reactions.   (2)

    iii What conditions would be used to bring about Reaction 2?   (2)

    iv What conditions would be used to bring about Reaction 3?   (2)

  c Name products **B** and **C**.   (2)

  d Propan-1-ol, **B** and **C** can be distinguished using their infrared spectra. State where the distinguishing absorptions would occur, and indicate the bonds responsible for the absorption(s) for each compound.   (3)

(Adapted from Chemistry (Nuffield) Q5, Jan 96)

Halogenoalkanes

A simple **halogenoalkane** consists of an alkane with one of the hydrogen atoms replaced by a halogen atom.
- The halogen part is named first; chloro-, bromo- or iodo-, as appropriate.
- This is followed by the name of the corresponding alkane.
- The carbon bonded to the halogen is numbered in the usual way (see page 45) to give the lowest possible number, and this number prefixes the name of the halogen.

As with alcohols, the **primary**, **secondary** and **tertiary** classification refers to the number of carbon atoms bonded to the carbon that is bonded to the halogen – so $CH_3CH_2CHBrCH_3$ (2-bromobutane) is a secondary halogenoalkane.

Preparation of halogenoalkanes

Halogenoalkanes are usually prepared by reaction of an alcohol with a hydrogen halide. The halogen from the halide **substitutes** for the OH group in the alcohol. There are two possible ways in which this can be done – either indirectly for bromoalkane or iodoalkane compounds, or directly for chloroalkanes.
- To form bromoalkanes or iodoalkanes, an alcohol reacts with the appropriate hydrogen halide formed by reaction of potassium bromide (or potassium iodide) with concentrated sulfuric acid. The halogenoalkane is distilled from the reaction mixture and collected under ice-cold water. Using concentrated sulfuric acid does not work very well for preparing iodoalkanes, because sulfuric acid oxidizes iodide ions to iodine and produces hardly any hydrogen iodide. It is better to react an alcohol with phosphorus(III) chloride.
- To form chloroalkanes, an alcohol is reacted with concentrated hydrochloric acid followed by shaking for about 20 minutes – the liquid chloroalkane forms as a separate layer.

The product can be purified by the steps below for purifying any organic liquid, like a halogenoalkane:
- **Add** anhydrous calcium chloride. (This step is only needed to prepare chloroalkane.)
- **Shake** in a separating funnel – allow to settle and discard the aqueous layer.
- **Wash** with sodium hydrogencarbonate solution – to react with any residual hydrogen halide – and separate again, keeping the organic layer.
- **Repeat washing** with sodium hydrogencarbonate until no more carbon dioxide is given off.
- **Dry** by shaking with anhydrous sodium sulfate.
- **Filter** through glass wool.
- **Redistil** collecting the liquid at the boiling temperature of the expected chloroalkane. This checks the **purity** of the product.

Reactions of the halogenoalkanes

There are three reactions you need to learn and classify:
- Reaction with aqueous alkali, such as potassium hydroxide, to form an alcohol.

$$CH_3CH_2CH_2CH_2Br(l) + KOH(aq) \rightarrow CH_3CH_2CH_2CH_2OH(aq) + KBr(aq)$$
      1-bromobutane   butan-1-ol

- Reflux/heat with concentrated alcoholic alkali to form an alkene.

$$CH_3CH_2CH_2CH_2Br + KOH(alc) \rightarrow CH_3CH_2CH\!=\!CH_2 + KBr + H_2O$$
      1-bromobutane   but-1-ene

- Heat under pressure with alcoholic ammonia solution (ammonia dissolved in ethanol), to form an amine.

$$CH_3CH_2CH_2CH_2Br + 2NH_3(alc) \rightarrow CH_3CH_2CH_2CH_2NH_2 + NH_4Br$$
      1-bromobutane   butylamine

- The first and last reactions above are examples of **nucleophilic substitution** reactions (see page 104).

ResultsPlus
Examiner tip ☑

You should be able to draw and label the apparatus for preparation of a halogenoalkane from an alcohol (including distillation) – it is examined in the Unit 2 paper, and also in Unit 3 Activity d. You also need to learn the steps for purifying an organic liquid.

Although you can make bromoalkanes from potassium bromide and sulfuric acid, there is a likelihood that some of the HBr will be oxidized by the sulfuric acid to bromine and so the yield will be reduced.

? Quick Questions

1   Draw structural formulae for the four isomers with molecular formula C_4H_9Br. State with reasons the order of reactivity of these compounds with hot aqueous silver nitrate.

2   An unknown halogenoalkane, with molecular formula C_2H_5X reacts with hot aqueous silver nitrate to give a yellow precipitate.
  a   Identify the halogen.
  b   Give the formula of the organic product of the reaction.
  c   State the type of organic reaction and the mechanism occurring.

3   When 2-bromobutane reacts with hot ethanolic potassium hydroxide three products form.
  a   Give the names for these products.
  b   State the type and mechanism of the reaction occurring in their formation.

- A **nucleophile** is literally a 'positive lover' – a species like a hydroxide ion, OH^-, with a negative charge; or a non-bonding pair of electrons in water or ammonia – which is ready to form a covalent bond with an electron-deficient carbon atom in the halogenoalkane. The hydroxide ion is the nucleophile in the first reaction; ammonia is the nucleophile in the last.
- The second reaction above, to form an alkene, is an **elimination** reaction (see page 104).

The carbon–halogen bond in halogenoalkanes is strongly polar. This is why nucleophiles are attracted to the δ– carbon atom and can form new bonds in halogenoalkanes

You might find it easier to remember the reactions of halogenoalkanes in the form of the spider diagram shown. You will also need structural, displayed and skeletal formulae and the names of all the reactants and products, along with the conditions required. You need to be able to interpret data and observations for the reactions of primary, secondary or tertiary halogenoalkanes.

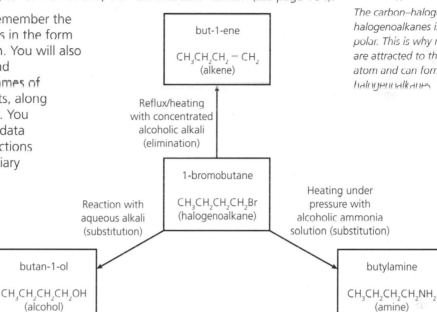

Reactions of halogenoalkanes

Rates of reaction of halogenoalkanes

The reaction used to investigate this is that with hot aqueous silver nitrate solution. The time taken for the silver halide precipitate to first appear indicates the rate of a reaction, and hence the reactivity of the carbon–halogen bond.
There are two factors to consider in the **reactivity**.
- the halogen present
- the number of **methyl groups** attached to the carbon bonded to the halogen.

The order of reactivity is that comparable fluoro- compounds are slowest through to the iodo- compounds which are fastest. The reason is that the *weakest* carbon–halogen bond breaks more easily and *not* the most polar bond, as you might guess. The most polar bond, $C-F$, is actually the strongest.

The reaction rate also increases as the number of methyl groups attached increases. In examinations, it is sufficient to remember that the reason for this is that attached methyl groups weaken the carbon–halogen bond.

Uses of halogenoalkane compounds

Because halogenoalkanes, like alkanes, are generally unreactive and, unlike alkanes, do not burn easily they are used as **fire retardants** and as **refrigerants**.

The **chlorofluorocarbons (CFCs)** used commonly in the past were found to seriously deplete the **ozone layer** in the atmosphere. This depletion enables more harmful **UV radiation** to reach the surface of the Earth, which increases the risk of skin cancer and eye cataracts.

Because they are unreactive, the CFCs are eventually carried through to the ozone layer, which they affect by a series of **free-radical reactions** (see page 106). This is why CFCs are being phased out – other halogenoalkanes, such as hydrofluorocarbons (HFCs) that do not release free radicals, are being used as safer alternatives.

Bond	Average bond enthalpy/kJ mol^{-1}
$C-F$	467
$C-Cl$	346
$C-Br$	290
$C-I$	228

Bond enthalpies for carbon–halogen bonds – the strength of the $C-F$ bond makes fluorocarbons inert and stable

Reaction mechanisms

You need to be able to **classify** organic reactions as addition, elimination, substitution, oxidation, reduction, hydrolysis or polymerisation and to define these terms.

Addition

An **addition reaction** is where two or more substances react to form one product. For example, when hydrogen bromide adds to propene, 2-bromopropane is formed:

$$CH_3CH{=}CH_2 + HBr \rightarrow CH_3CHBrCH_3$$

Elimination

In an **elimination reaction**, one reactant breaks down to produce a new substance and a small molecule, such as water or hydrogen bromide, is eliminated. For example, when 2-bromopropane reacts with hot, alcoholic potassium hydroxide to form propene. The hydrogen bromide eliminated reacts with the potassium hydroxide to form potassium bromide and water.

$$CH_3CHBrCH_3 + KOH \rightarrow CH_3CH{=}CH_2 + KBr + H_2O$$

We say that hydrogen bromide is eliminated – even though it goes on to react with something else.

Substitution

In a **substitution reaction**, one atom or group of atoms is replaced by another atom or group of atoms. For example when 1-chlorobutane is made from butan-1-ol, a chlorine atom is 'swapped' for the OH group (see page 97):

$$CH_3CH_2CH_2CH_2OH(l) + HCl(aq) \rightarrow CH_3CH_2CH_2CH_2Cl(l) + H_2O(l)$$

In the substitution reaction of halogenoalkanes with aqueous alkali or with ammonia (previous page), a halide ion is swapped for either a hydroxide ion or the amino group, NH_2.

Oxidation and reduction

In organic chemistry, it is helpful to use the old definitions – oxidation is addition of oxygen or removal of hydrogen; reduction is addition of hydrogen or removal of oxygen. For example, the oxidation of propan-2-ol to propanone involves the removal of two hydrogen atoms:

$$CH_3CH(OH)CH_3 + [O] \rightarrow CH_3COCH_3 + H_2O$$

Hydrolysis

Hydrolysis occurs when a molecule is broken down by the addition of water. This may involve water, acid or alkali as the inorganic reactant. The reaction of 2-bromopropane with aqueous alkali to form propan-2-ol is a hydrolysis reaction:

$$CH_3CHBrCH_3(l) + OH^-(aq) \rightarrow CH_3CHOHCH_3(aq) + Br^-(aq)$$

Polymerisation

Polymerisation happens when many identical small molecules join to form a long chain. For example, when propene forms poly(propene):

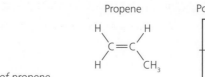

Polymerisation of propene

The letter 'n' represents a large number of molecules. Note that there are now no double bonds in the polymer structure.

Bond breaking

This can be **homolytic** or **heterolytic**.

- **Homolytic bond breaking** occurs when each atom involved in the bond keeps one of the shared electron pair in the bond that breaks. This forms two atoms, or molecules, each with an unpaired electron.

  For example, in the **initiation** step of the **free-radical substitution** reaction between methane and chlorine (see Unit 1, page 58), two chlorine radicals form:

  $$Cl_2 \rightarrow Cl\cdot + Cl\cdot$$

- **Heterolytic bond breaking** occurs when both of the electrons in a bond are kept by one of the atoms involved, forming a positive ion and a negative ion.

  For example, in the dissociation of HCl into H^+ and Cl^- ions:

  $$HCl \rightarrow H^+ + :Cl^-$$

Cl $\cdot \cdot$ Cl $\rightarrow$ Cl$\cdot$ + Cl$\cdot$

Homolytic fission of Cl_2 – the half-headed arrows each show the movement of one electron

Cl — Cl $\rightarrow$ Cl$^+$ + Cl$^-$

H — Cl $\rightarrow$ H$^+$ + Cl$^-$

Heterolytic fission of Cl_2 and of HCl – the full curly arrows represent the movement of a pair of electrons

Classification of reagents

- A **free radical** is an atom or molecule with an unpaired electron. For example, Cl· in the free-radical substitution of an alkane with chlorine.
- An **electrophile** is an atom, molecule or ion that accepts a pair of electrons to form a covalent bond. For example, bromine in electrophilic addition to an alkene (see Unit 1, page 61).
- A **nucleophile** is an atom, molecule or ion that donates a pair of electrons to form a new covalent bond. For example, water when it substitutes for a halogen atom in the hydrolysis of a halogenoalkane.

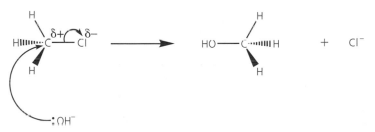

The nucleophile attacks the back of the carbon atom donating an electron pair. This causes a new bond to form and the carbon halogen bond to break.

Nucleophilic substitution reaction mechanism for the reaction of chloromethane with aqueous hydroxide – the curly arrow shows the movement of a pair of electrons

Predicting the type of mechanism

- Polar bonds will always break heterolytically.
- Non-polar bonds, such as carbon–hydrogen bonds, will usually break homolytically, but not always.
- A nucleophile can attack the δ+ atom in a polar bond.
- An electrophile can attack an electron-rich part of a molecule – for example, the C=C (π) bond in alkenes.

?) Quick Questions

1 Classify the following reactions:
   a $CH_3CH_3 + Br_2 \rightarrow CH_3CH_2Br + HBr$
   b $CH_2{=}CH_2 + Br_2 \rightarrow CH_2BrCH_2Br$
   c $CH_3CH_2Br + H_2O \rightarrow CH_3CH_2OH + HBr$
   d $CH_3CH_2Br \rightarrow CH_2{=}CH_2 + HBr$
2 Classify the type of attacking reagent in the first three reactions in question **1**.

Green chemistry

1 Suggest two ways in which an industrial plant can reduce use of energy for reaction mixtures that need to be heated.

2 Suggest two reasons why the use of catalysts is economically beneficial.

3 Why is ethanoic acid not produced industrially by the method you used in the laboratory?

Reinvention of processes in the chemical industry

This is being done to make the processes more sustainable or 'greener' in five ways:
- Changing to renewable resources.
- Finding alternatives to hazardous chemicals.
- Developing new catalysts with higher **atom economies**, reducing unwanted products (see page 15).
- Making more efficient use of energy.
- Reducing waste and preventing pollution.

Use of CFCs and HCFCs

The use of CFCs as refrigerants and fire retardants is being phased out due to their harmful effect on the ozone layer. They are being replaced with HFCs (see page 103).

CFCs and free-radical reactions in the ozone layer

Ozone (O_3) is produced from oxygen in the ozone layer of the stratosphere. Oxygen molecules break up to form oxygen free radicals, due to absorbing **ultraviolet radiation (UV)** from the Sun. The oxygen free radicals formed then recombine with oxygen to form ozone:

$$O=O \xrightarrow{UV} O· + O·$$
$$O_2 + O· \rightarrow O_3$$

Ozone protects the Earth from harmful UV, but is broken down by reactions with CFCs and also oxides of nitrogen (NO and NO_2) produced by car and jet engines.

CFCs are very stable molecules, but high in the stratosphere they are dissociated by UV radiation to form highly reactive chlorine free radicals:

$$CCl_2F_2 \rightarrow ·CClF_2 + Cl·$$

The chlorine free radical formed reacts with ozone:

$$Cl· + O_3 \rightarrow ·ClO + O_2$$

The chlorate free radical formed reacts with ozone producing more chlorine free radicals:

$$·ClO + O_3 \rightarrow 2O_2 + Cl·$$

Because the Cl· is regenerated and unchanged, it is a **catalytical radical**.

The net effect is to convert two molecules of ozone to three molecules of oxygen. Note that a chain reaction is set up in which a chlorine free radical reacts to form another chlorine free radical – so just one molecule can do an incredible amount of damage.

Nitrogen oxides also damage the ozone layer by producing NO· radicals which react with ozone:

$$NO·(g) + O_3(g) \rightarrow NO_2(g) + O_2(g)$$

Each NO· is then regenerated by the NO_2 reacting with O atoms:

$$NO_2(g) + O·(g) \rightarrow NO·(g) + O_2(g)$$

The NO· is regenerated and is unchanged when it breaks down ozone.

Greenhouse gases and global warming

The effects of greenhouse gases

Infrared radiation (IR) from the Sun, which has short wavelength, mostly passes through the atmosphere and is absorbed by the Earth's surface. This heats the Earth, which re-emits longer wavelength IR. Any **greenhouse gases** in the atmosphere **effectively reflect** this longer wavelength IR, warming the atmosphere.

The *relative* greenhouse effect of a gas varies because molecules with different bonds absorb IR differently. The **global warming potential** of a gas combines its ability to absorb IR with its lifetime in the atmosphere. The concentration of a gas in the atmosphere also affects its potential to cause warming.
- Carbon dioxide has a low global warming potential, but levels are increasing.
- CFCs such as trichlorofluoromethane have a much higher global warming potential, but overall concentrations are very low.

Anthropogenic and natural climate changes

- **Anthropogenic climate change** results from human activities such as burning fossil fuels or deforestation that increase levels of carbon dioxide, methane and dinitrogen oxide (N_2O, from aircraft exhausts) over relatively short timescales.
- **Natural climate change** is due to natural processes such as the dissolving of carbon dioxide in sea water or the formation of carbonates in rocks that remove carbon dioxide from the atmosphere over much longer timescales. Other natural processes, such as volcanic eruptions or changes in solar activity, can also cause climate change.

Carbon neutrality and carbon footprint

A **carbon-neutral fuel** is one for which the release of carbon dioxide in its manufacturing and its burning equals the absorption of carbon dioxide from the atmosphere as the raw material is grown or the fuel formed. Only certain **biofuels** can be considered to be carbon neutral within a human lifetime. **Fossil fuels** are formed on geological timescales.
- A **carbon-neutral process** occurs when there is no *net* carbon dioxide emission to the atmosphere. Emissions are balanced by actions that remove an equivalent amount of carbon dioxide.

The **carbon footprint** of a fuel is the total mass of carbon dioxide produced from a fuel as it is manufactured and then burnt, in units of $g\,kJ^{-1}$. For biofuels such as those made from sugar or vegetable oils, the amount of carbon dioxide that was absorbed in growing the raw material for the fuel should be subtracted from this total mass.
- A carbon footprint *in general* is a measure of the amount of carbon dioxide emitted through the use of fossil fuels. It is often measured in tonnes of carbon dioxide, and can be calculated for an individual, a household, an organisation or over a product **lifecycle** for manufactured goods.

> ### Thinking Task
>
> Compare the environmental damage caused by travelling 500 miles in a petrol-driven car with flying 500 miles in a jet aeroplane.

> ### Quick Questions
>
> 1 Explain why carbon dioxide contributes more to the increased greenhouse effect than methane, even though methane has a higher greenhouse warming potential.
> 2 What is the difference between *natural* and *anthropogenic* change?
> 3 Explain why some gases, such as oxygen, have a zero global warming factor.
> 4 Give reasons why these fuels may be considered as not being carbon neutral:
>   a hydrogen made from reforming methane
>   b a biofuel such as bioethanol.

Topic 5: Organic chemistry 2 – halogenoalkanes and green chemistry checklist

By the end of this topic you should be able to:

Revision spread	Checkpoints	Specification section	Revised	Practice exam questions
Halogenoalkanes	Name halogenoalkanes, and draw their displayed, structural and skeletal formulae. Classify them as primary, secondary or tertiary	2.10.2a	☐	☐
	Interpret and explain the effect of primary, secondary and tertiary structures on reactivity	2.10.2b	☐	☐
	Prepare halogenoalkanes from alcohols, and purify organic liquids like halogenoalkanes	2.10.2c	☐	☐
	Describe the reactions of halogenoalkanes with: aqueous alkali, hot alcoholic alkali, water containing silver nitrate, alcoholic ammonia	2.10.2d and e	☐	☐
	Describe the uses of halogenoalkanes	2.10.2f	☐	☐
Reaction mechanisms	Classify reactions as addition, elimination, substitution, oxidation, reduction, hydrolysis or polymerization	2.11a	☐	☐
	Show that you understand that homolytic bond breaking forms free radicals, heterolytic bond breaking involves nucleophiles or electrophiles	2.11b	☐	☐
	Define the terms free radical, nucleophile and electrophile	2.11c	☐	☐
	Say why it is helpful to classify reagents	2.11d	☐	☐
	Use the link between bond polarity and the type of reaction mechanism for a compound	2.11e	☐	☐
	Describe the mechanisms of substitution reactions of alkanes, alkenes and halogenoalkanes	2.11f	☐	☐
	Describe how the ozone layer absorbs UV and how CFCs and nitrogen oxides deplete the ozone layer, including chemical equations for the appropriate free-radical reactions	2.11g	☐	☐
Green chemistry	Describe how chemical processes are being made more sustainable by: (i) changing to renewable resources (ii) finding alternatives to hazardous chemicals (iii) finding catalysts with higher atom economies (iv) making more efficient use of energy (v) reducing waste and preventing pollution	2.13a	☐	☐
Greenhouse gases and global warming	Discuss the relative effects of different gases on global warming	2.13b	☐	☐
	Describe the difference between anthropogenic and natural climate changes	2.13c	☐	☐
	State the definitions of carbon neutrality and carbon footprint	2.13d	☐	☐
	Apply the principles of carbon neutrality to different fuels	2.13e	☐	☐
	Explain why CFCs are no longer used, including the detail and mechanism of their reaction with the ozone layer	2.13f	☐	☐

ResultsPlus
Build Better Answers

1 a i Describe how the halogenoalkane 2-chloro-2-methylpropane can be prepared in the laboratory. (2)
  ii How would you separate 2-chloro-2-methylpropane from the preparation mixture? It is less dense than water and aqueous solutions. (2)
  iii Why would you wash the product with sodium hydrogencarbonate solution? How would you know when it had been adequately washed with sodium hydrogencarbonate solution? (2)
  iv How would you dry your 2-chloro-2-methylpropane before the final distillation step? (2)

  b 2-chloro-2-methylpropane reacts rapidly with hot silver nitrate solution. 1-chlorobutane reacts very slowly with hot silver nitrate solution.

    i Describe what you would see during and at the end of each reaction if the mixture was left in sunlight.   (3)

    ii Explain why the reaction with 2-chloro-2-methylpropane is so much faster than with 1-chlorobutane.   (2)

☑ **Examiner tip**

You should have carried out the particular reaction described here. If not, you should be able to deduce an appropriate method from your own preparation of a named halogenoalkane. You should also be able to describe steps in the purification of a liquid organic product.

a **i** Shake 2-methylpropan-1-ol (1) with concentrated hydrochloric acid. (1) A maximum of (1) for stating 'a substitution reaction of an alcohol with concentrated hydrochloric acid'.

    ii Use a separating funnel (1) discard the lower aqueous layer/keep the upper layer. (1)

    iii To neutralize traces of hydrochloric aid that may not have dissolved in the aqueous layer (1) until no bubbles of gas (carbon dioxide) can be seen. (1)

    iv Add anhydrous sodium sulfate; (1) filter through glass wool (accept decant). (1)

b **i** A white (1) precipitate; (1) darkens in sunlight. (1).

    ii Because the tertiary compound (1) has the weaker C—Cl bond. (1)

    ◼ **Basic answer:** Most students can recall the tests for the halide ions and the characteristic colours. However, some write 'cloudy' instead of precipitate, and lose marks.

    ⚠ **Excellent answer:** To get full marks for the explanation, students need to describe the **relative strength** of the C–halogen bonds in primary, secondary and tertiary halogenoalkanes (tertiary halogenoalkanes have weaker C–halogen bonds than primary halogenoalkanes). Note that answers describing the different strength of the carbon–halogen bond in chloro- or iodo-halogenoalkanes would not gain any marks.

(From Chemistry (Nuffield) CN 1 Q5, Jan 96)

Practice exam question

1 This question is about the hydrolysis of halogenoalkanes.

- 2 cm³ of ethanol is added to each of three test tubes.
- Three drops of 1-chlorobutane are added to the first, three drops of 1-bromobutane are added to the second, and three drops of 1-iodobutane are added to the third test tube.
- A 2 cm³ portion of hot, aqueous silver nitrate solution is added to each test tube.
- A precipitate forms immediately in the third test tube, slowly in the second test tube, and extremely slowly in the first test tube.
- In each reaction the precipitate is formed by silver ions, $Ag^+(aq)$, reacting with the halide ions formed by hydrolysis of halogenoalkanes.

  a **i** Why was ethanol added to each test tube?   (1)

    ii The same organic product forms in each reaction. Name this organic product.   (1)

    iii Complete the equation for the hydrolysis of 1-bromobutane.   (1)

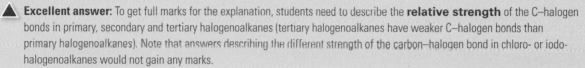

$$C_4H_9Br + H_2O \rightarrow$$

    iv What is the colour of the precipitate in the third test tube?   (1)

    v Name the precipitate that forms slowly in the first test tube and write an ionic equation, including state symbols, for its formation.   (3)

    vi Ammonia is added to the precipitate formed in the first test tube. Describe and explain what you would observe.   (2)

    vii Explain why the rates of hydrolysis of the three halogenoalkanes are different.   (2)

  b 1-bromobutane reacts with an alcoholic solution of potassium hydroxide at high temperatures to form but-1-ene.

    i Draw a fully labelled diagram to show the apparatus for carrying out this reaction in the laboratory.   (3)

    ii Suggest a chemical test for an alkene such as but-1-ene. State the colour change you would observe.   (2)

  c Explain why chlorofluorocarbons (CFCs) should not be used in aerosols or fire retardants, but fluorocarbons are quite acceptable.   (5)

(Adapted from Chemistry (Nuffield) Q2, June 96)

Unit 2: Practice unit test

Section A

1 How many of the following molecules will absorb IR radiation?

CO, CO_2, NO, NO_2, N_2, O_2

A two   **B** three   **C** four   **D** five   (1)

2 Which intermolecular forces exist between molecules of propanal, CH_3CH_2CHO?
  A instantaneous–induced dipole only
  B permanent dipole–permanent dipole only
  C instantaneous dipole–induced dipole and hydrogen bonds
  D instantaneous dipole–induced dipole and permanent dipole–permanent dipole (1)

3 Which of these metals will give a yellow flame colour?
  A calcium   **B** potassium   **C** sodium   **D** magnesium   (1)

4 Which of the molecules PCl_3, CO, CO_2 and CCl_4 are polar?
  A all four   **B** PCl_3 and CO
  C CO and CCl_4 **D** PCl_3 and CO_2 (1)

5 A chlorine atom is also often described as
  A a nucleophile   **B** an electron   **C** an electrophile   **D** a free radical   (1)

6 What is the oxidation number of nitrogen in dinitrogen difluoride, N_2F_2?
  A -1 **B** -3 **C** $+1$ **D** $+3$ (1)

7 Which of these compounds is least soluble in water?
  A barium hydroxide   **B** calcium hydroxide
  C magnesium hydroxide   **D** strontium hydroxide   (1)

8 Which of these statements about fluorine is **not** correct?
  A It is a gaseous element at room temperature and pressure.
  B It can react with chloride ions to form chlorine.
  C It forms salts with Group 1 metals.
  D It is less electronegative than chlorine. (1)

9 The following liquids have the same number of electrons per molecule. Which is likely to have the lowest boiling temperature?
  A $CH_3CH_2CH_2CH_2OH$ **B** $CH_3CH_2CH_2CH_3$
  C $CH_3C(CH_3)_3$ **D** $CH_3CH(CH_3)CH_2CH_3$ (1)

10 What would be the effect of increasing the concentration of $Ag^+(aq)$ ions in the reaction:

$$Fe^{2+}(aq) + Ag^+(aq) \rightleftharpoons Fe^{3+}(aq) + Ag(s)?$$

  A Rate of reaction increases, yield of $Fe^{3+}(aq)$ remains the same.
  B Rate of reaction increases, yield of $Fe^{3+}(aq)$ decreases.
  C Rate of reaction decreases, yield of $Fe^{3+}(aq)$ remains the same.
  D Rate of reaction increases, yield of $Fe^{3+}(aq)$ increases. (1)

11 Which of these is likely to be the best solvent for pentan-1-ol?
  A $H_2O(l)$ **B** $CH_3COCH_3(l)$
  C $NaCl(aq)$ **D** $CH_3CH_2CH_2CH_2CH_2CH_3(l)$ (1)

12 By considering intermolecular forces, suggest which of these liquids flows most slowly when poured.

    A propan-1,2,3-triol   **B** propan-1,2-diol

    C pentane   **D** butane   (1)

13 Four organic compounds are:

    A butanoic acid   **B** butanal

    C butanone   **D** but-1-ene

    a Identify the product of vigorous oxidation of butan-1-ol.   (1)

    b Identify the product of oxidation of butan-2-ol.   (1)

14 Which of these is a primary alcohol?

    A 3-methylpentan-2-ol   **B** pentan-2-ol

    C pentan-3-ol   **D** 2-methylpentan-1-ol   (1)

15 Which of these reactions is **not** a disproportionation reaction?

    A $2H_2O_2(aq) \rightarrow O_2(g) + 2H_2O(l)$

    B $S_2O_3^{2-}(aq) + 2H^+(aq) \rightarrow SO_2(g) + S(s) + H_2O(l)$

    C $Cl_2(aq) + 2Br^-(aq) \rightarrow 2Cl^-(aq) + Br_2(aq)$

    D $2Cu^+(aq) \rightarrow Cu(s) + Cu^{2+}(aq)$ (1)

16 $10.0\,cm^3$ of $0.05\,mol\,dm^{-3}$ potassium hydroxide solution was placed in a conical flask and titrated with $0.100\,mol\,dm^{-3}$ hydrochloric acid solution, using methyl orange indicator.

    a What would be the colour of phenolphthalein at the end point?

        A pink   **B** colourless   **C** yellow   **D** orange   (1)

    b Where would you put the hydrochloric acid solution?

        A burette   **B** pipette

        C volumetric flask   **D** measuring cylinder   (1)

    c What volume, in cm^3, of $0.100\,mol\,dm^{-3}$ acid was added at the end point?

        A 5.00   **B** 10.00   **C** 15.00   **D** 20.00   (1)

17 Molecules absorb IR radiation because

    A they change their polarity when they vibrate.

    B they change their velocity when they vibrate.

    C they change their magnetic field when they vibrate.

    D they change their direction of rotation when they vibrate   (1)

Section B

18 Bromine is extracted from sea water using chlorine.

    a **i** Write the ionic equation for the reaction between chlorine and bromide ions.   (1)

        ii The sea water is acidified before the addition of the chlorine to prevent the bromine produced reacting with the water:

$$Br_2(aq) + H_2O(l) \rightleftharpoons HOBr(aq) + HBr(aq)$$

        Name the type of reaction taking place between bromine and water. Explain your answer in terms of the oxidation numbers of bromine.   (3)

       iii Bromine vapour reacts with sulfur dioxide and water as follows:

$$Br_2(g) + SO_2(g) + 2H_2O(l) \rightarrow 2HBr(aq) + H_2SO_4(aq)$$

        State the oxidation numbers of sulfur in SO_2 and H_2SO_4.   (2)

       iv Use the data from **(iii)** to show that bromine is acting as an oxidizing agent.   (1)

b The ionic equation for the reduction of iodate(V) ions to iodine in acid is:

$$2IO_3^-(aq) + 12H^+(aq) + 10e^- \rightarrow I_2(aq) + 6H_2O(l)$$

    i Write the ionic half-equation for the oxidation of SO_2 in water to SO_4^{2-}.   (1)

    ii Combine the reduction reaction of iodate(V) ions, IO_3^-, with the oxidation reaction of SO_2 to give the full ionic equation for the reaction of IO_3^- with SO_2.   (2)

(From Edexcel AS Chemistry Unit 1 Q5, June 08)

19 Phosphorus reacts with a limited amount of chlorine to form phosphorus trichloride.

  a **i** Draw a dot-and-cross diagram to show the arrangement of electrons in phosphorus trichloride, PCl_3. You need only show outer electrons.   (2)

      ii Draw the phosphorus trichloride molecule, making its shape clear and giving the bond angle.   (1)

  b Explain the shape of the phosphorus trichloride molecule, and why the bond angle is different from the bond angle in methane.   (3)

(From Edexcel AS Chemistry Unit 1 Q7, June 08)

20 This question is about the reactions and properties of some halogenoalkanes.

  a State the reagents and conditions needed to convert the following halogenoalkanes into the named product:

    i 1-bromobutane into butan-1-ol

    ii 1-iodobutane into butylamine

    iii 2-chloropropane into propene.   (4)

  b Two gaseous halogenoalkanes that could be used as fire retardants have structural formulae CF_2ClBr and CF_3CHF_2.

    i Give the systematic name of CF_2ClBr.   (1)

    ii Suggest TWO reasons to explain how these compounds can help put out fires.   (2)

    iii Explain why fire retardants containing some halogenoalkanes, such as CF_2ClBr, are being phased out. Suggest a reason why the scientific community still supports the use of fire retardants containing CF_3CHF_2.   (4)

(Adapted from Edexcel AS Chemistry Unit 2 Q18, Jan 09)

21 Some water companies use chlorine to purify water for domestic use. Concentrations of chlorine are carefully monitored by testing water samples.

  a Excess potassium iodide was added to a $1000\,cm^3$ sample of water. The iodine formed reacted with $14.0\,cm^3$ of $0.0100\,mol\,dm^{-3}$ sodium thiosulfate solution.

    i Calculate the number of moles of sodium thiosulfate, $Na_2S_2O_3$, used in the reaction.   (1)

    ii Copy and complete the ionic equation for the reaction between iodine and sodium thiosulfate.   (2)

$$I_2(aq) + 2S_2O_3^{2-}(aq) \rightarrow$$

    iii Calculate the number of moles of iodine molecules used in the reaction.   (1)

    iv Write the ionic equation for the reaction between chlorine molecules and iodide ions.   (2)

    v Write down the number of moles of chlorine in the sample.   (1)

    vi Calculate the mass of chlorine in the sample.   (1)

    vii The maximum accepted concentration of chlorine in drinking water is 0.5 parts per million. Show by calculation that the sample of water is acceptable.

You may assume $1000\,cm^3$ of water has mass $1000\,g$.   (2)

  b Suggest **two** reasons why the concentration of chlorine in domestic water must not exceed 0.5 parts per million.   (2)

  c The whole of the solution containing chlorine was used in one titration. Explain how this affects the reliability of your answer.   (2)

(Adapted from Edexcel AS Chemistry,
Sample assessment material for Unit 2 Q28, Sept 07)

22 **a** Draw a diagram to show the distribution of molecular energies in a reaction at two temperatures T_1 and T_2, where T_2 is higher than T_1.   (3)

   b Explain how a catalyst speeds up the rate of a reaction.   (3)

   (Amended from Edexcel AS Chemistry Unit 2 Q20, Jan 09)

Section C

Quality of written communication will be tested in questions marked with an asterisk.

23 Ethanoic acid is used industrially in the manufacture of polymers and glues, and also in the food industry as an acidity regulator.

   It can be synthesized in the laboratory by the reaction of ethanol with excess sodium dichromate(VI) solution when it is acidified with concentrated sulfuric acid. Ethanol is placed in a suitable flask along with some antibumping beads. The concentrated sulfuric acid is added a drop at a time. The sodium dichromate(VI) solution is then added a drop at a time, causing the mixture to boil spontaneously. When the addition of the sodium dichromate(VI) solution is complete, the mixture is heated under reflux for approximately 15 minutes. The ethanoic acid formed can then be separated from the reaction mixture.

   Ethanoic acid can also be produced by the Cativa™ process. Methanol, which can be obtained from wood, is reacted with carbon dioxide at 20–30 atmospheres and 190°C in the presence of an iridium catalyst:

   $$CH_3OH(g) + CO(g) \rightleftharpoons CH_3COOH(l)$$

   a i Why are the concentrated sulfuric acid and the sodium dichromate(VI) solution added a drop at a time?   (1)

   ii Draw a labelled diagram of the apparatus that could be used to heat the mixture under reflux.   (3)

   iii What colour would the mixture be after it was heated under reflux?   (1)

   b A solution containing both water and ethanoic acid is produced by distillation of the final reaction mixture.

   i Explain why the other products and any excess reactants are left behind in the distillation flask.   (1)

   ii Suggest a method to separate pure ethanoic acid, boiling temperature 118°C, from the water.   (1)

   c i In the Cativa™ process what effect, if any, would increasing the pressure have on the yield of ethanoic acid? Justify your answer.   (2)

   ii Suggest **two** reasons why it might be difficult, or undesirable, to produce ethanoic acid in industry by scaling up the laboratory process.   (2)

 *d An alternative industrial process for the production of ethanoic acid is the oxidation of butane using a transition metal catalyst at 150°C and 55–60 atmospheres.

   $$2C_4H_{10}(l) + 5O_2(g) \rightarrow 4CH_3COOH(aq) + 2H_2O(l)$$

   Evaluate the 'greenness' of the two industrial processes.

   Suggest **two** further pieces of information that would help you to make a more informed decision.   (6)

   (From Edexcel AS Chemistry Q21, Jan 09)

Chemistry Laboratory Skills I

Your laboratory skills will be tested by four practical activities based on the content of Units 1 and 2. There are 40 marks available – 20% of the total AS marks.

Activity a: General practical competence (GPC)

At the end of the AS course, your teacher will verify that you have developed practical skills by carrying out a range of core practicals, listing five of them on your Edexcel record sheet. Marks are not given for this activity.

For Activities **b**, **c** and **d** you will carry out practical tasks under controlled conditions. You are not allowed to use any notes or books. The tasks are set by Edexcel, and you will not know in advance what the task will be. At least one task must be completed for each of Activities **b**, **c** and **d**. If more than one is completed, the task scoring the highest mark is included in your total mark for Unit 3.

Activity b: Qualitative observation (14 marks)

In this activity you will carry out reactions on a test-tube scale. You will record your observations and use these to identify unknown compounds and to explain the reactions you have seen.

Chemical reactions and tests that may be included in this activity include:
- Heating Group 1 and 2 nitrates and carbonates (see page 81).
- Flame tests for Group 1 and 2 compounds (page 81).
- The barium chloride test for the sulfate ion (page 81).
- The test for halide ions in solution by the addition of aqueous silver nitrate followed by ammonia solution (see page 87).
- The reaction between solid potassium halides and concentrated sulfuric acid (see page 87).
- The test for Group 2 ions in solution by the addition of sodium hydroxide solution (page 81).
- The test for the ammonium ion by warming an ammonium salt with sodium hydroxide solution to liberate ammonia (see below).
- The test for alkenes by the addition of bromine water or potassium manganate(VII) (page 58).
- The reaction of sodium and of phosphorus(V) chloride with alcohols (page 96).
- Warming a halogenoalkane with aqueous silver nitrate (page 103).
- Warming primary and secondary alcohols with acidified potassium dichromate(VI) (page 97).
- Tests for oxygen and carbon dioxide (from GCSE), and for ammonia (colourless gas, turns damp red litmus paper blue) and hydrogen chloride (steamy fumes, turns damp blue litmus paper red) gases.

Record your observations using appropriate chemical terms:
- On mixing two solutions a **precipitate** may form – avoid using 'suspension' or 'solid'.
- If a **gas** is given off from a solution, record observations as 'bubbles' or 'effervescence'.
- When a silver halide precipitate dissolves in ammonia solution, a **colourless** not 'clear' solution is formed.
- **Steamy fumes** (not 'white smoke') of hydrogen chloride are evolved when concentrated sulfuric acid is added to solid potassium chloride. When these fumes mix with ammonia **white smoke** is seen.

You should be able to make inferences (deductions) from these tests and write equations for the reactions. For example, if steamy fumes are given off when PCl_5 is added to a liquid, the inference is that hydrogen chloride is being evolved and that an OH group is present.

ResultsPlus
Examiner tip

Take notice of the marks available for each test. Generally one mark is awarded for an observation of each change, so if two marks are available there will be two changes to observe and record.

You may also be given a mass spectrum and IR spectroscopic data to help you to identify an organic compound.

ResultsPlus
Build Better Answers

In an assessment exercise for Activity b, students are given an inorganic compound **X**. They are told to:
a carry out a flame test on **X**
b add dilute nitric acid and aqueous silver nitrate followed by concentrated ammonia solution to a solution of **X**.
The students are told to identify **X** by giving its formula.
In basic answers, students record a yellow flame in **(a)** and that an 'insoluble yellow precipitate' is formed in **(b)**. They identify **X** as NaI.
In excellent answers, students record that, on careful observation, there were traces of a lilac colour in the flame in **(a)**. They record that in
(b) the pale yellow precipitate formed did not dissolve in concentrated ammonia solution. They identify **X** correctly as KI.

Quick Questions

1 Give the colours of the lithium, sodium, potassium, calcium and barium flames.
2 What is observed when aqueous silver nitrate is added to a solution of:
  a sodium chloride
  b potassium bromide
  c sodium iodide?
3 An organic compound with 3 carbon atoms in its molecule:
  • reacts with sodium to give off bubbles
  • turns acidified potassium dichromate(VI) from orange to green on warming
  • has a peak at $m/z = 29$ in its mass spectrum.
Identify the compound by giving its structural formula.

ResultsPlus
Watch out!

If you are asked to identify a compound by writing its formula then giving its name will not be accepted as an answer.

Activity c: Quantitative measurement (14 marks)

In this task you will carry out experiments requiring measurements of volume, mass and temperature. These will include using a thermometer, burette, pipette, balance, measuring cylinder and volumetric flask.

If necessary you will be allowed to answer the questions and draw any graphs in a separate session, but you will not be allowed to take your results away with you at the end of the first session.

You will carry out at least one of these assessment tasks:
• Acid–base titration – finding the molar mass of a solid acid (see pages 82–83).
• Finding the enthalpy change for the reaction between an acid and a base (see page 20).
• Sodium thiosulfate–iodine titration (see page 88).
• Hess's law (see page 21).

Marks are awarded for:
• reading and recording measurements accurately and precisely
• carrying out calculations using your results and graph
• evaluating the experiment.

Accuracy and precision in measurements
• A volumetric pipette measures one volume only – normally 25.0 cm³ or 10.0 cm³. When draining a pipette, a small volume of solution should remain in the tip – the calibration includes this small volume, so if you shake or blow it out the volume will be inaccurate.

- Read a pipette or burette by lining up the bottom of the meniscus with the graduation at eye level.
- A burette volume should be recorded to the nearest 0.05 cm³ – for example, 23.45 cm³, 24.60 cm³.
- You should repeat titrations in an attempt to obtain two concordant titres to within 0.20 cm³, and then average these to 2 decimal places.
- To make a standard solution, after you dissolve a weighed sample of a solid in water you use a volumetric flask to make up the solution to an exact volume – normally 250 cm³ or 100 cm³.
- Record temperatures to the degree of precision asked for in the instructions. If you are told to record the temperature of a solution to the nearest 0.5°C and it is exactly 35°C then record this as 35.0°C.

Calculations

- You are not allowed any notes or books in the assessments, so you will have to remember the methods used to calculate concentrations of solutions, molar masses, energy transferred and enthalpy changes.
- When you calculate the energy transferred in a reaction include units, either J or kJ, with your answer.
- When you calculate an enthalpy change, ΔH, give your answer in kJ mol^{-1} and include a sign in front of the value.

 Quick Questions

1   Choose titres from the following to calculate a mean titre: 23.45, 23.80, 23.35 cm³.
2   Give the colour change at the end-point in a titration in which sodium hydroxide solution is added to hydrochloric acid with phenolphthalein as indicator.
3   Calculate the energy transferred, in kJ, in a reaction in which the temperature of 50 cm³ of solution falls by 8.6°C. The specific heat capacity of the solution is 4.2 J g^{-1} K^{-1}.

 Thinking Task

24.50 cm³ of 0.100 mol dm^{-3} sodium hydroxide solution exactly neutralized 25.0 cm³ of a solution containing 2.37 g of an acid, HX in 250 cm³ of solution.

$$NaOH(aq) + HX(aq) \rightarrow NaX(aq) + H_2O(l)$$

Calculate the molar mass of HX.

Evaluating

The analysis and evaluation skills from practical work in Units 1 and 2 will show you the type of errors to look out for, the limitations of the techniques used, accuracy of measurement and uncertainty of your final result.

- Remember that repeating an experiment using the same apparatus will not improve the accuracy, because the same systematic errors could be present.
- You should be able to discuss possible reasons for any differences between your result and the accepted or true value. For example, the most likely cause of an inaccurate result in a titration is improper use of a burette or pipette.

Activity d: Preparation (12 marks)

You will be assessed by carrying out a preparation using laboratory apparatus that will be familiar to you. You will either record the mass of your product or carry out a simple test on it, and then answer questions on the procedure. For this activity you may work in a pair with another student – but you must answer the questions individually.

You will carry out at least one of these preparations:
- Preparation of a double salt (see page 14).
- Preparation of a salt (see page 14).
- Preparation of an organic compound (see pages 96–97, 102).

Laboratory practices that you may need to use in the preparation

- Controlled heating using a Bunsen burner.
- Use of a balance reading to at least 0.1 g.
- Measuring volume with a measuring cylinder.
- Adding one reagent to another, making sure that the reaction does not become too vigorous.
- Filtration.
- Evaporating a solution to reduce its volume.
- Drying crystals using filter paper.
- Assembling glassware for organic preparations.

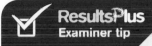

ResultsPlus
Examiner tip

You may be asked to explain the purpose of steps in the procedure you are given. For example, to explain why reagents must be mixed slowly and carefully – so that the reaction does not become too vigorous.

Calculating maximum mass and yields

- Normally one of the reagents in a preparation is in excess. Given an equation for the reaction, you can calculate the theoretical maximum mass of product using the quantity of the reagent that is not in excess (see page 14).
- The yield of product will be lower than the maximum mass. You may be asked to calculate the percentage yield (see page 14).
- You should be able to suggest why a percentage yield is lower than 100%.

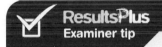

ResultsPlus
Examiner tip

Reasons for a low yield differ from one preparation to another – they include an incomplete reaction, some product remaining in solution after crystallization, side-reactions and impure reagents. Make sure that you have some evidence for your suggested reason.

Quick Questions

1   In the preparation of crystals of hydrated copper(II) sulfate, copper(II) carbonate is added to $50 \, cm^3$ of warm $1.0 \, mol \, dm^{-3}$ sulfuric acid until there is no further reaction.

$$CuCO_3(s) + H_2SO_4(aq) \rightarrow CuSO_4(aq) + H_2O(l) + CO_2(g)$$

   a How would you know when the reaction had finished?
   b Calculate the maximum mass of hydrated copper(II) sulfate, $CuSO_4 \cdot 5H_2O$, that may be obtained from the solution formed. The molar mass of $CuSO_4 \cdot 5H_2O$ is $250 \, g \, mol^{-1}$.
   c If the actual yield of hydrated copper(II) sulfate is 4.5 g calculate the percentage yield.

Answers

Answers to quick questions

Chemical quantities and formulae

TT ^{24}Mg may react slightly slower than ^{23}Mg because it is heavier, but because it is the (outer) electrons, not the neutrons or protons, that take part in chemical reactions, they will be very much the same.

1 **a i**   0.500 mol which has 3.010×10^{23} atoms;
    ii 55.56 mol which has 3.344×10^{25} atoms
   b 23 mmol (2 s.f. only)
2 A_r for Si = 28.1
3 **a** NaClO   **b i** $AlCl_3$;   **ii** Al_2Cl_6

Chemical equations and reacting masses

1 **a** $KBrO_3(s) + 3C(s) \rightarrow KBr(s) + 3CO(g)$
   b $2ZnS(s) + 3O_2(g) \rightarrow 2ZnO(s) + 2SO_2(g)$
2 **a** $BaCl_2(aq) + Na_2SO_4(aq) \rightarrow BaSO_4(s) + 2NaCl(aq)$
      $Ba^{2+}(aq) + SO_4^{2-}(aq) \rightarrow BaSO_4(s)$
   b $2HNO_3(aq) + Ca(OH)_2(aq) \rightarrow Ca(NO_3)_2(aq) + H_2O(l)$
      $2H^+(aq) + 2OH^-(aq) \rightarrow 2H_2O(l)$
3 0.01 mol Zn produces 0.02 mol Ag so
   $Zn(s) + 2AgNO_3(aq) \rightarrow Ag(s) + Zn(NO_3)_2(aq)$

Reactions with gases

1 **a** $13.5 \, dm^3$ **b** $500 \, cm^3$
2 $0.120 \, dm^3$ ($120 \, cm^3$)
3 $4NH_3(g) + 5O_2(g) \rightarrow 4NO(g) + 6H_2O(g)$
TT The volume of a gas changes with temperature. 0° Celsius = 273.15 kelvin. The Kelvin scale is used in thermodynamics because gas volume is then directly proportional to temperature ($V_1T_1 = V_2T_2$).

Percentage yield and atom economy

1 **a** 3.30 g (3 s.f.)
   b 94.5%; not all the salt crystallized from the solution; some product left on filter paper when drying the crystals
2 75.1%

Enthalpy changes and enthalpy level diagrams

1 As on page 18 for the endothermic diagram, with 'reactants' for 'ice' and 'products' for 'melted ice'.
2 As on page 18 for the exothermic diagram, with 'supersaturated solution' for 'reactants' and 'solid' for 'products'.
3 As on page 18 for the endothermic diagram, but with the enthalpy change labelled as +15.8 kJ.
4 Products are at a lower energy level than the reactants, hence $\Delta H = H_{products} - H_{reactants}$ is negative.
TT Heat energy from the hot water is absorbed and changes the solid into the supersaturated solution at a higher energy level. It remains at this level, even on cooling, so long as the solution is not shocked.

Measuring enthalpy changes

1 All the energy produced is transferred to the solutions; there is no heat absorbed by the calorimeter; there is no heat loss to the room or from the surface of the solutions; the specific heat capacity of the solutions is the same as that of water.

2 5.1 kJ. NB no need to calculate amounts of acid and alkali, and alkali is in excess.
3 $-1444 \, kJ \, mol^{-1}$
TT A little different since energy is required to dissociate the molecular acid as the H^+ ions are neutralized.

Using Hess's law

1 **a** 2.1 kJ,   **b** $85 \, kJ \, mol^{-1}$
2 **a** 0.25 kJ,   **b** $10 \, kJ \, mol^{-1}$
3 $-94 \, kJ \, mol^{-1}$

Bond enthalpy

1 **a** $-42 \, kJ \, mol^{-1}$
   b The substances are not in their standard states.
2 Bond enthalpies: $F_2(g) < Cl_2(g) > Br_2(g) > I_2(g)$. Bond enthalpies are a guide to, but do not explain, the relative reactivity of the halogens; also have to consider bond length and electronegativity.
TT The reaction is exothermic, graphite being the more (energetically) stable. It does not tell us anything about kinetic stability – how much energy is needed to get the reaction to proceed. Diamonds are not turning into graphite!

Mass spectrometry

1 Each molecule has a unique mass spectrum.
2 **a** CO_2 = 44, NaOH = 40, $CaCO_3$ = 100, CH_3OCH_3 = 46, Na_2SO_4 = 142, CH_3OH = 30.
   b Those for CO_2, CH_3OCH_3 and CH_3OH.
3 Similarity: same atomic number – the same number of protons. Difference: different numbers of neutrons.
4 32.10
5 44
TT Assumptions: relative abundance of ^{13}C has been constant over the ages; the half-life of ^{13}C is constant.

Ionization energy and electron shells

1 $738 + 1451 = 2189 \, kJ \, mol^{-1}$.
2 $1s^2 2s^2 2p^6 3s^2 3p^6 3d^{10} 4s^2 4p^3$
3 Full or half-full sub-shells are relatively stable.
4 Big jump between 2nd and 3rd ionization energies; the 2 electrons lost easily must be in the outer shell, the 3rd electron is in an inner shell and is harder to remove.
TT Zn is $1s^2 2s^2 2p^6 3s^2 3p^6 3d^{10} 4s^2$;
   Zn^{2+} is $1s^2 2s^2 2p^6 3s^2 3p^6 3d^{10}$
   This now has a full outer shell, which is very stable, even though it does not have a rare gas configuration.

Electronic configurations and periodic properties

1 They are the ones that take part in chemical reactions.
2 1st I.E. of Na > K because the outer electron in Na is closer to the nucleus; it also has 1 shell fewer and so less shielding from the attraction, therefore more difficult to remove.
   1st I.E. of Na < Ar because the outer electron in Ar is more attracted to the more positive nucleus, therefore more difficult to remove.

3 P has simple molecular structure (P_4). Little energy is required to overcome the intermolecular forces, thus allowing relatively free movement of P_4 molecules. Diamond has a giant molecular structure. Each C atom is surrounded by 4 others, all covalently bonded. Much more energy is required to break the strong covalent bonds and allow free movement of C atoms.

Ionic bonding

1 All directions.

2 By atoms gaining or losing one or more electrons.

3

$$\left[\text{Mg}\right]^{2+} \begin{bmatrix} \!\!\!\times\!\!\times \\ \!\!\times\!\! \text{Cl} \!\times\!\! \\ \!\!\times\!\!\times \end{bmatrix}^{-} $$
$$\begin{bmatrix} \!\!\!\times\!\!\times \\ \!\!\times\!\! \text{Cl} \!\times\!\! \\ \!\!\times\!\!\times \end{bmatrix}^{-}$$

4 **a** N^{2-}(0.171 nm), O^{2-}(0.140 nm), F^{-}(0.133 nm), Na^{+}(0.102 nm), Mg^{2+}(0.072 nm), Al^{3+}(0.053 nm)

 b Electronic configurations are all the same (isoelectronic) – $1s^2 2s^2 2p^6$.

 c Because electrons in excess of the proton number are held less tightly to the nucleus; where there are fewer electrons than the proton number they are attracted more strongly to the nucleus and are drawn closer to it.

Lattice energies and Born–Haber cycles

1 Add up the 1st, 2nd and 3rd ionization energies for aluminium.

2 Ionic bonding is strong, because lattice energies are very large.

Testing the ionic model

1 Charge density – i.e. ionic radius and charge. The higher the charge density, the greater the polarizing power.

2 Because the electrons are further from the nucleus than in a small ion; therefore they are held less tightly. There may well be more inner shells providing shielding, further reducing the hold of the nucleus on the electron cloud.

3 By being distorted, creating a region of electron density between the ions

4 The lattice enthalpy is significantly more exothermic than the value predicted from theoretical considerations, and the melting temperature is lower than that expected of ionic compounds.

TT When it is not omnidirectional, there is some electron density between the ions, the cation has polarized the anion – the ions are not spherical, the cation is small and/or has a high charge, the anion is large.

TT When there is more electron sharing than electron transfer, electron density is more or less evenly distributed between the bonded atoms.

Covalent bonding

1 The shared electron pair forms a region of high electron density between the two atoms which attracts the nuclei of each atom.

2

Cl_2O

$c \equiv o$

Al_2Cl_6

3 It is very hard, with high melting temperature so covalent bonds are strong, requiring a lot of energy to break them apart or to allow the atoms to move as a liquid; it does not conduct electricity since all the electrons are used in bonding, there is none free to move.

TT Covalent bonds vary:
- 1, 2 or 3 shared electron pairs
- electrons donated to the bond by both or just one of the atoms
- polarity of bonds due to different electronegativities of the bonded atoms
- enthalpy of formation varies, as does bond length.

Metallic bonding

1 To melt, the particles in the solid have to be able to move past each other, i.e. have to have sufficient energy to overcome the forces of attraction between them. In metals, these forces of attraction are metallic bonds, which require a lot of energy to break.

2 Conduction of heat requires transmission of kinetic energy. 'Hot' delocalized electrons (greater kinetic energy) are free to collide with their 'cooler' neighbours and pass on kinetic energy.

3 The positive ions in a metal lattice are held in place by the attraction of delocalized electrons all around them.

TT Group 1 metals have a lower charge density than transition metals and donate one electron each to the sea of electrons. Transition metals have ions with high charge density and donate at least two electrons to the sea of electrons – hence large forces of attraction requiring lots of energy to break. The attractive forces between ions and electrons is greater if the ions have a higher charge density and there are more electrons.

Hazard and risk in organic chemistry

1 A risky experiment is one in which the hazards are not kept to a minimum, the risk of harm is too great; a hazardous chemical is a material with measured harmful effects.

2 Diluting it to less than $0.5 \, mol \, dm^{-3}$

3 Keeping the quantity to a minimum, using it in a fume cupboard, wearing gloves of the correct protective standard, having a 'spill kit' to hand, e.g. thiosulfate solution and absorbent granules such as vermiculite.

TT It is very carcinogenic and there are good alternative materials readily available that present a much lower risk.

Organic compounds and functional groups

1   **a** C_6H_{14} **b** C_8H_{16} **c** $C_5H_{11}OH$ **d** C_3H_7I
2   The hydroxyl group, —OH.
3   The chemical reactions are the same.
TT Carbon atoms are smaller than silicon – they have one fewer electron shell. Therefore, there is less screening of the nucleus and less electron cloud repulsion. So the bond length is shorter, the two nuclei are closer to the electron cloud between them, and the attraction is greater.

Naming and drawing organic compounds

1 CH_2OHCH_2OH
2   Tetrachloromethane (CCl_4)
TT A regular hexagon with one of the sides showing a double bond.

Hydrocarbons: alkanes and alkenes

1   Pentane, 2-methylbutane and 2,2-dimethylpropane.
2   Separates crude oil into fractions, which are mixtures of hydrocarbons of varying boiling temperatures.
3   Produces shorter-chain molecules from long-chain molecules that are of less use.
TT Advantages: clean burn; available from water; therefore renewable resource. Disadvantages: has to be stored and transported under pressure; energy is used to extract it from water or methane, methane is a non-renewable resource; it slowly leaks (diffuses) through its storage vessels.

Naming geometric isomers

1   No *cis-trans* for any of them.
    a   *E*-1,2-dichloroprop-1-ene
    b   *Z*-1,2-dichloroprop-1-ene
    c   *E*-1-bromo-1-chloro-2-iodoethene
    d   *Z*-1-bromo-1-chloro-2-iodoethene.
2   In their physical properties such as melting and boiling temperatures.
TT When there are different functional groups or side chains attached to the carbon atoms.

Reactions of alkanes

1   A species with an unpaired electron.
2   To break the covalent bond in Cl_2 molecule.
3   Different termination steps produce different products; there can be further substitution of other H atoms by more Cl atoms.
4   Produces carbon monoxide which is an invisible, odourless and toxic gas (can be fatal).
TT By repeated substitution:
$CH_4 + Cl_2 \rightarrow CH_3Cl + HCl$
$CH_3Cl + Cl_2 \rightarrow CH_2Cl_2 + HCl$
$CH_2Cl_2 + Cl_2 \rightarrow CHCl_3 + HCl$
$CHCl_3 + Cl_2 \rightarrow CCl_4 + HCl$

Reactions of alkenes

1   200°C and Ni catalyst (finely divided).
2   1,2-dichloroethane, CH_2ClCH_2Cl.
3   an electron-seeking species, it is attracted to centres of high electron density.
TT The structure that gives rise to their colour is destroyed – e.g. Br_2 is brown because it contains the Br—Br bond; this is destroyed when it adds across the double bond.

Polymers

1   Full combustion requires very high temperatures, using fossil fuel resources and increasing CO_2 emissions; any carbon in the waste will be converted to CO_2, which increases the greenhouse effect.
2

poly(propene)

3   Energy and materials used, CO_2 emissions produced during extraction of raw materials, original manufacture, recycling, reuse or disposal.

Shapes of molecules and ions

1   Tetrahedral, with bond angle 107°; four pairs of electrons including one non-bonding pair; adopt minimum repulsion, tetrahedral arrangement; non- bonding pairs repel more than bonding pairs, so reduced bond angle.
2   **a**

    b

Note there are three bonding pairs and one non-bonding pair. The shape of H_3O^+ is the same as that of ammonia.
3   Four bonding pairs take up minimum repulsion arrangement of tetrahedron.

Bond polarity and intermediate bonding

1   **a** $C^{\delta+}$—$Cl^{\delta-}$ The bond is polar because chlorine is much more electronegative than carbon. The molecule is polar because it is asymmetric.
    b $C^{\delta+}$—$Br^{\delta-}$ The bond is polar because bromine is more electronegative than carbon. The molecule is non-polar because it is perfectly symmetrical.
    c   C—H bond is non-polar because the atoms have similar electronegativities. The molecule is non-polar.
    d $C^{\delta+}$=$O^{\delta-}$ The bond is polar because oxygen is much more electronegative than carbon. The molecule is non-polar because it is perfectly symmetrical.

Intermolecular forces

1   **a**   2-methylpropane < butane < 1-bromo-2-methylpropane < 1-bromobutane.
    b   1-bromobutane is highest due to largest number of electrons, no branching and therefore the largest London forces.
       1-bromo-2-methylpropane has the same number of electrons as butane, but is unbranched so the molecules cannot align so well.
       Butane has fewer electrons and 2-methylpropane has the same number of electrons as butane but is unbranched.

2 Both have similar London forces because both have the same number of electrons, but ethanol has additional hydrogen bonds between the hydrogen attached to the oxygen and the oxygen atom of other molecules.

3 a No **b** No **c** No **d** Yes, both between H of HF and O of water, and between H of water and F of HF

4

bond angle O—H---O is 180°

Hydrogen bonds; note that the three atoms associated in a hydrogen bond always form an angle of 180°.

Solubility

1 Stains that can't be removed by water contain molecules with high London forces between them, which are too strong to be broken by water molecules. Ethanol has a carbon chain which can form London forces with water molecules, which makes it soluble in water.

2 1,1,1-trichloroethane molecules are held to each other by dipole–dipole forces and London forces; water molecules are held together by hydrogen bonds. 1,1,1-trichloroethane does not form strong attractions to water molecules, so the energy required to overcome the hydrogen bonds cannot be supplied by the energy released by the new forces of attraction between water and 1,1,1-trichloroethane. However, both hexane and tetrachloromethane are held to each other by similarly weak forces, so these can be overcome, and similarly weakly attractive forces can be made between the two liquids.

Oxidation and reduction

1 N_2O: +1, dinitrogen oxide; HNO_3: +5, nitric acid; $NaNO_2$: +3, sodium nitrate(III); $(NH_4)_2SO_4$: −3, ammonium sulfate.

2 a copper(II) oxide
 b copper(I) oxide
 c chromium(III) sulfate
 d chromium(II) sulfate

Redox reactions

1 a Sulfur in thiosulfate changes from +2 to 0 in sulfur, and +4 in sulfur dioxide.
 b Sulfur has been oxidized and reduced in the same reaction.

2 $MnO_4^-(aq) + 8H^+(aq) + 5Fe^{2+}(aq) \rightarrow Mn^{2+}(aq) + 5Fe^{3+}(aq) + 4H_2O(l)$

Properties and reactions of Group 2 elements and compounds

1 a The white solid is calcium oxide, the solution is calcium hydroxide, and the white precipitate is calcium carbonate.
 b $CaO(s) + H_2O(l) \rightarrow Ca(OH)_2(aq)$
 $Ca(OH)_2(aq) + CO_2(g) \rightarrow CaCO_3(s) + H_2O(l)$

2 Crush rock with a pestle and mortar; dip a nichrome wire into concentrated hydrochloric acid and the crushed rock; put the wire in a Bunsen flame.

Acid–base titrations

1 $0.0500\,\text{mol}\,\text{dm}^{-3}$
2 1.30%
3 a Readings 1 and 2, because the titres (23.90 and 23.80) are within $0.20\,\text{cm}^3$ of each other, while reading 3 (24.10) is not.
 b Error in reading the burette; titration run too fast.

Group 7 and reactions of the halides

1 a $Br_2(aq) + 2I^-(aq) \rightarrow 2Br^-(aq) + I_2(s)$
 b $Br_2(aq) + 2KOH(aq) \rightarrow KOBr(aq) + KBr(aq) + H_2O(l)$
 Ionic equation is $Br_2(aq) + 2OH^-(aq) \rightarrow OBr^-(aq) + Br^-(aq) + H_2O(l)$

2 $NH_3(g) + HI(g) \rightarrow NH_4I(s)$; white smoke seen.
 $2HI(g) \rightarrow H_2(g) + I_2(s)$; purple gas changes to a grey/black solid.

Iodine–thiosulfate titrations

1 $I_2(s) + 6KOH(aq) \rightarrow KIO_3(aq) + 5KI(aq) + 3H_2O(l)$
 Add iodine to boiling concentrated potassium hydroxide solution until it is just coloured. Add one drop of potassium hydroxide solution. Allow to cool and suction filter.

2 A

Reaction rates and catalysts

1 a There are more molecules in a given volume, so they collide more frequently, and the rate of a reaction increases.
 b Strong containers are needed which is expensive; to maintain high pressure, energy is needed which is expensive.

2 The peak of the curve moves to the right and is lower; the proportion of molecules with energy greater or equal to the activation energy increases dramatically.

Chemical equilibria

1 An equilibrium is set up: $Cl_2(g) + ICl(l) \rightleftharpoons ICl_3(s)$
 Pouring off chlorine gas causes the reaction to go in the direction that makes more chlorine (by Le Chatelier's principle), so solid iodine trichloride disappears and liquid iodine monochloride forms.

2 Low temperature is preferred because the reaction is exothermic; high pressure causes the reaction to go in the forward direction which produces fewer gaseous molecules.

Alcohols

1 Butan-1-ol and 2-methylpropan-1-ol.
2 a $C_2H_5OH(l) + Na(s) \rightarrow C_2H_5ONa(s) + \frac{1}{2}H_2(g)$
 b Sodium ethoxide
3 a HCHO, methanal; HCOOH, methanoic acid
 b HCHO gives a red precipitate with boiling Benedict's solution; HCOOH neutralizes sodium carbonate solution with effervescence.

Mass spectra and IR absorption

1 Ethanol would give a broad peak at $3600–3200\,\text{cm}^{-1}$ due to the hydrogen-bonded O—H bond; ethanal would give a sharp peak at about $1740\,\text{cm}^{-1}$ due to the C=O bond; ethanoic acid would give a broad peak between 3500 and $2500\,\text{cm}^{-1}$ due to the extensively hydrogen-bonded O—H bond.

2 Both would have peaks at the same highest m/z value, corresponding to the parent ion. There would be different minor peaks due to different masses of different fragment ions, because the structures fragment in different ways.

3 For example, OH^+ (from the OH group); 17

4 a 58 due to parent ion $CH_3CH_2CHO^+$; 57 due to $CH_3CH_2CO^+$; 29 due to $CH_3CH_2^+$.
   b Peak at 58 due to the (same) parent ion; no peak at 29 because no CH_3CH_2 group; peak at 43 due to CH_3CO^+; peak at 15 due to CH_3^+.

5 Nitrogen dioxide and ammonia are weak greenhouse gases; chloromethane and chlorine are potentially greenhouse gases; argon is not.

Halogenoalkanes

1 $CH_3CH_2CH_2CH_2Br$, $(CH_3)_2CHCH_2Br$, $CH_3CH_2CHBrCH_3$, $(CH_3)_3CBr$; are the isomers in order of increasing reactivity; as the branching increases the carbon–bromine bond becomes weaker, so the reaction is faster.

2 a Iodine is the halogen.
   b The organic product is C_2H_5OH.
   c The reaction is a substitution reaction with initial attack by a nucleophile.

3 a But-1-ene, E-but-2-ene and Z-but-2-ene.
   b They are all formed by elimination of hydrogen bromide.

Reaction mechanisms

1 a substitution   **b** addition
   c substitution   **d** elimination

2 Free radical, electrophile and nucleophile.

Green chemistry

1 Insulation of part of the apparatus; reduce energy demand of reaction by using a catalyst that works at lower temperatures. Note, microwave ovens work well for heating on a laboratory scale but not yet on an industrial scale.

2 Lower temperatures so less energy required; better yield; new catalysts improve atom economy and so reduce cost of disposal of waste product.

3 Oxidation of ethanol using acidified sodium dichromate(VI) produces a mixture of products, including ethanal, so the process has a low atom economy. Direct formation from methanol and carbon monoxide has an atom economy of almost 100%.

Greenhouse gases and global warming

1 Much more carbon dioxide is produced by burning fossil fuels, than methane (mainly produced by cows).

2 Anthropogenic changes are changes caused by the activities of human beings, such as burning fossil fuels or deforestation.
   Natural changes are due to natural processes like dissolving of carbon dioxide in sea water, or the formation of carbonates in rocks.

3 Oxygen does not change its dipole when it vibrates, so does not absorb or 'reflect' IR, so it is not a greenhouse gas and does not lead to global warming.

4 a The feedstock may be from a renewable and carbon-neutral source (biomass, not fossil fuel) but the industrial process uses energy, which is most likely generated from burning fossil fuels.
   b Carbon emissions (energy produced from fossil fuels) from production of the fertiliser, pesticides and tools used for growing the crops, processing and refining the crops, transport and infrastructure (particularly for export).

Activity b: Qualitative observation

1 red, yellow, lilac, red, green
2 a White precipitate   **b** Cream precipitate
   c (Pale) yellow precipitate
3 $CH_3CH_2CH_2OH$

Activity c: Quantitative measurement

1 23.40 cm^3 **2** colourless to pale pink
3 1.8 kJ
TT 96.7 g mol^{-1}

Activity d: Preparation

1 a Effervescence ends; some copper(II) carbonate remains undissolved.
   b 12.5 g **c** 36%

Answers to practice exam questions: Unit 1

Topic 1

1 a The simplest ratio (1) of atoms of each element present in a compound (1)
   b masses divided by A_r (1), ratio 1:1:1.49 (1), smallest whole ratio leads to $Na_2S_2O_3$ (1)

2 a $M_r \, Na_2CO_3 = 106$; amount of $Na_2CO_3 = \frac{100}{106} = 0.943 \text{ mol}$ (1) amount of $NaHCO_3 = 2 \times 0.934 = 1.89 \text{ mol}$ (1) $M_r \, NaHCO_3 = 84$, so mass $NaHCO_3 = 1.89 \times 84 = 159 \text{ g}$ (1)
   b 0.943 mol (1) so volume $CO_2 = 0.934 \times 24 = 22.4 \text{ dm}^3$ (1)

3 a Percentage of Al in $Al_2O_3 = 52.9\%$ (1), so 1 tonne Al_2O_3 contains 0.529 tonnes (529 kg) aluminium (1)
   b $90\% \times 0.529 \text{ t} = 0.476 \text{ t Al}$ (1); to obtain 1 t Al requires $\frac{1}{0.476} \text{ t} = 2.10 \text{ t } Al_2O_3$ (1)

4 a Amount = volume × concentration $= \frac{250}{1000} \times 0.0470 = 0.0118 \text{ mol}$ (1) $M_r(H_2C_2O_4.2H_2O) = 126$ (1) Mass = M_r × amount = $126 \times 0.0118 = 1.49 \text{ g}$ (1)
   b Amount = 250/1000 × 0.050 = 0.0125 mol (1); $M_r \, Na_2CO_3 = 106$ (1); mass = $106 \times 0.0125 = 1.325 \text{ g}$ (1)

5 **B** (1)   **6** **C** (1)   **7** **D** (1)

Topic 2

1 Total energy bonds broken = (6 C—H) + (2 C—C) + (1 C=O) + (4 O=O) = $+5969 \text{ kJ mol}^{-1}$ (3)
   Total energy bonds formed = (6 C=O) + (6 H—O) = $-7614 \text{ kJ mol}^{-1}$ (2)
   $\therefore \Delta H_c = (+5969 - 7614)$ (1) $= -1645 \text{ kJ mol}^{-1}$ (1)

2 $\Delta H_f^\ominus = \Delta H_1 - \Delta H_2$
$\Delta H_1 = 3 \times \Delta H_c[C] + 4 \times \Delta H_c[H_2]$ (1) −
−2326 kJ mol^{-1} (1)
$\Delta H_f^\ominus = -2326 - (-2219) = -107$ kJ mol^{-1} (1)

Topic 3

1 $S^- < S < S^+$ **2 C**
3 a $(1s^2)2s^22p^63s^23p^64s^2$
  b i Energy/enthalpy/heat energy change required
     per **mole** (1) for the **removal of 1 electron**
     (1) from **gaseous atoms** (1) *not* molecules
     or $X(g) \rightarrow X^+(g) + e^-$ (states required for 2nd
     and 3rd marks) (2)
    ii (Even though) there is a greater nuclear
     charge/number of protons (1) the **outer/
     valency** electron(s) are further from nucleus
     (1) and are **more** shielded *or* have **more**
     (filled) inner shells/electrons (1).
4 a Vapour bombarded with high energy electrons
    (1) to knock off electrons (1).
  b An electric field (1) **c** A magnetic field (1)
5 **C** (1)

Topic 4

1 **D** (1)
2 a Ionic bonding (1)
  b
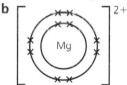 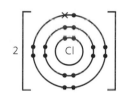
                                  (3)
  c Metallic (1)
  d Mg^{2+} ions are smaller than Na^+ ions (1). Two
    electrons lost per Mg^{2+} ion, only one per Na^+ ion
    (1). More energy needed to overcome stronger
    forces of attraction between Mg^{2+} ions than Na^+
    ions (1).
3 a

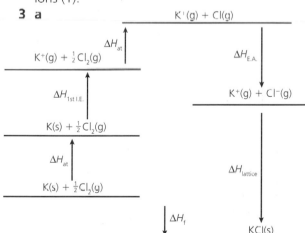

    (2) for correct order in diagram.
    $\Delta H_f = \Delta H_{at}[K] + \Delta H_{1stIE}[K] + \Delta H_{at}[C] + \Delta H_{EA}[Cl]$
    $+ \Delta H_{latt}[KCl]$ (1)
    Electron affinity of $Cl(g) = (-436.7) - (+89.2) -$
    $(+121.7) - (+419) - (-711)$ (1)
    $= -355.6$ (kJ mol^{-1}) (1)
  b Increased electrostatic attraction due to higher
    charge of Ca^{2+} ion relative to K^+ ion (1).
    Smaller size of Ca^{2+} ion relative to K^+ ion (1).
    [Ca^{2+}ion greater charge density than K^+ ion gets
    max. 1]

c The calculated value is based on the assumption
   that the solid is 100% ionic (1).
   KCl is only very slightly covalent, so little
   difference (1).
   $CaCl_2$ is more covalent because the cation is more
   polarizing (higher charge, smaller radius) (1).

Topic 5

1 a
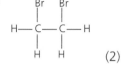
                            (2)
  b Bromine solution added to unknown (1). Red/
    orange colour of bromine (1). Goes colourless (1)
  c Repeating unit of CH_2, no double bonds shown (2)
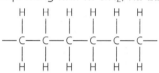
2 a i Free-radical substitution (1)
    ii 1 mark for $Cl-Cl \rightarrow Cl\cdot + Cl\cdot$ (1) for correct
     curly arrows
    iii $C_2H_6 + Cl\cdot \rightarrow \cdot C_2H_5 + HCl$ (1)
     $\cdot C_2H_5 + Cl_2 \rightarrow C_2H_5Cl + Cl\cdot$ (1)
    iv Either $\cdot C_2H_5 + \cdot C_2H_5 \rightarrow C_4H_{10}$
     or $\cdot C_2H_5 + Cl\cdot \rightarrow C_2H_5Cl$ (1)
  b i Electrophilic addition (1)
    ii $C_2H_4 + Br_2 \rightarrow CH_2Br-CH_2Br$ (1)
    iii The brownish-reddish colour (1) of the
     bromine water will decolorise (1).

Answers to practice exam questions: Unit 2

Topic 1

1 **B** (1)   **2** **A** (1)   **3** **A** (1)   **4** **D** (1)
5 a
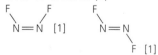
                      BF_3
    (1) mark for each correct diagram and (1) mark
    for each correct bond angle.
  b i $\overset{\times\times}{\underset{\times\times}{\times}}F\overset{\bullet\bullet}{\underset{\bullet\bullet}{\times}}N\overset{\bullet}{\underset{\bullet}{\bullet}}N\overset{\bullet\bullet}{\underset{\bullet\bullet}{\times}}F\overset{\times\times}{\underset{\times\times}{\bullet}}$ (2)
    ii Correct sequence (1) electrons – can be all
     dots or crosses (1)

    F   F   F
    $\diagdown$ $\diagup$ $\diagdown$
     N═N [1]   N═N
                           $\diagdown$
                           F [1]

    iii Bond energy varies with environment (1).
  c i
    Two crosses or dots between N and B (1). Rest
    of the detail (1).
    ii Dative (1) covalent (1)

Topic 2

1 a $2Na(s) + \frac{1}{2}O_2(g) \rightarrow Na_2O(s)$

b $Sr(s) + \frac{1}{2}O_2(g) \rightarrow SrO(s)$

c $Ca(s) + 2H_2O(l) \rightarrow Ca(OH)_2(aq) + H_2(g)$

d $Mg(s) + H_2O(g) \rightarrow MgO(s) + H_2(g)$

e $2NaOH(aq) + H_2SO_4(aq) \rightarrow Na_2SO_4(aq) + 2H_2O(l)$

f $Ca(OH)_2(aq) + 2HCl(aq) \rightarrow CaCl_2(aq) + 2H_2O(l)$
Formulae and balancing (1); state symbols (1) for each part.

2 a Group 2 (1); large jump between second and third ionization energies (1).

b R; has the lowest first ionization energy (1). Electron is furthest from the nucleus/electron is more shielded from the nucleus/electron in a new quantum shell (1).

3 **B** (1)

4 **C** (1)

Topic 3

1 **B** (1)

2 **A** (1)

3 a i Yellow (1) precipitate (1)

ii $Ag^+(aq) + I^-(aq) \rightarrow AgI(s)$ formulae (1) state symbols if formulae correct (1)

b i Precipitate darkens (1)

ii Photography (1)

iii The precipitate dissolves to give a clear colourless solution (1) due to the formation of a more stable compound/ion (1).

c i Hydrogen bromide/HBr (1)

ii $HBr(g) + NH_3(g) \rightarrow NH_4Br(s)$ (1)

iii Bromine/Br_2 (1)

iv Sulfur dioxide/SO_2 (1)

v As an acid/proton donor (1) and as an oxidizing agent (1)

Topic 4

1 **D** (1)

2 **C** (1)

3 a i Sodium (1)

ii White (1) solid (1)

b i Oxidation/redox (1)

ii Sodium dichromate(VI) (1); sulfuric acid (1)

iii Dilute acid (1); distillation (1)

iv Concentrated acid and twice as much/excess oxidizing agent (1); reflux followed by distillation (1)

c **B** is propanal (1); **C** is propanoic acid (1)

d Propan-1-ol has a broad absorption due to O—H in an alcohol 3500–3000 cm^{-1} (1).
Propanal has an absorption due to C=O at about 1730 cm^{-1} (1).
Propanoic acid has a broad absorption due to O—H in a carboxylic acid at 3500–2500 cm^{-1} (1).

Topic 5

1 a i To act as a solvent/dissolves the halogenoalkane (1)

ii Butan-1-ol (1)

iii $C_4H_9Br + H_2O \rightarrow C_4H_9OH$ (1) + HBr (1)

iv Yellow (1)

v Silver chloride (1); $Ag^+(aq) + Cl^-(aq) \rightarrow AgCl(s)$ formulae (1) balancing and state symbols (1)

vi Precipitate dissolves (1) because a more stable product forms (1).

vii C—I bonds are weaker (and break more easily) (1) than C—Cl bonds (1).

b i Horizontal test tube containing ceramic fibre soaked in halogenoalkane and alcoholic potassium hydroxide (1); arrow with heat at ceramic fibre (1); gas collected over water (1).

ii Bromine water (1) from yellow to colourless (1) or acidified potassium manganate(VII) (1) from pink/purple to colourless (1)

c Any of the following to a maximum of five marks:
CFCs cause less ozone to absorb harmful UV (1)
CFCs decompose to form chlorine atoms/free radicals (1)
Each chlorine free radical causes the destruction of thousands of ozone molecules (1)
One of the equations: (1)
$Cl\cdot + O_3 \rightarrow ClO\cdot + O_2$
$ClO\cdot + O\cdot \rightarrow Cl\cdot + O_2$
Fluorocarbons do not break down to give free radicals (1) because the C—F bond is too strong (1).

Practice unit test answers: Unit 1

Section A

1	C (1)	2	B (1)
3	C (1)	4	A (1)
5	B (1)	6	B (1)
7	C (1)	8	D (1)
9	B (1)	10	A (1)
11	B (1)		**Total 11 marks**

Section B

12 a

b $Mg(s) + Cl_2(g) \rightarrow MgCl_2(s)$; formulae (1) state symbols (1) – only if formulae are correct or near miss for $MgCl_2$ (e.g. $MgCl/Mg_2Cl$)

c $\dfrac{(56.25 \times 70) + (37.50 \times 72) + (6.25 \times 74)}{100}$ (1)

$= 71$ (1)

d $\dfrac{4.73}{71}$ moles (1); $\times 30.6 = 2.04\,dm^3$ (1)
answer with no working 1 max.

e Metallic (1); attraction between Mg^{2+} (1) and (surrounding) sea of electrons/delocalized electrons (1)

f Ionic (1)

or

Correct charges and number of ions (1)
Correct electronic structures (1)

Total 14 marks

13 a The energy required to remove 1 mole of electrons (1) from 1 mole of atoms (1) in the gaseous state (1).

b $Na^+(g) \rightarrow Na^{2+}(g)$ (1) $+ e^-$ (1)

c The atomic radius decreases (1); outer (valence) electrons are closer to nucleus (1); therefore more strongly attracted to nucleus (1).

d A magnesium electron is removed from full s sub-shell, this is particularly stable (1). Therefore requires more energy than removing single p electron from Al (1).

Total 10 marks

14 a The number of atoms in 12 g (1) of ^{12}C (1)/the number of atoms in 1 mole of ^{12}C

b i Amount of $Z = \dfrac{2.87 \times 10^{22}}{6.02 \times 10^{23}}$ (1) $= (0.04767)$

$\dfrac{6.02 \times 10^{23}}{2.87 \times 10^{22}} \times 1.907$ is 1 mol

$= 40.(0)$ (1)

Allow 39.7 for 2 marks – this is rounding 0.04767 to 2 s.f. in calculation
Allow 38.14 for 1 mark – this is rounding to 1 s.f.

ii Ar/argon (1)

c i Amount of hydrogen peroxide produced

$= \dfrac{3.09\,g}{34\,g\,mol^{-1}} = 0.09088$ (moles) (1)

Amount of potassium superoxide
$= 0.09088 \times 2$ (moles) (1)
Mass of potassium superoxide
$= 0.09088 \times 2 \times 71 = 12.9\,g$ (1) or 13 g

ii Volume of oxygen $= \dfrac{3.09\,g}{34\,g} \times 24\,dm^3$

$= 2.18\,dm^3$ (1)

Total 9 marks

15 a C_nH_{2n+2} (1)

b Cracking (1)

c Skeletal (1)

d i C_9H_{20} (1)

ii 3-ethyl-4-methylhexane (1);
allow methyl before ethyl, 4-methyl-3-ethylhexane, 3-methyl-4-ethylhexane, 4-ethyl-3-methylhexane and 3,4-ethylmethylhexane

Total 5 marks
Total 38 marks

Practice unit test answers: Unit 2

Section A

1	C (1)	2	D (1)	3	C (1)	4	B (1)
5	D (1)	6	C (1)	7	C (1)	8	D (1)
9	C (1)	10	D (1)	11	B (1)	12	A (1)
13 a	A (1)	b	C (1)				
14	D (1)	15	C (1)				
16 a	B (1)	b	A (1)		c	A (1)	
17	A (1)					**Total 20 marks**	

Section B

18 a i $Cl_2(aq) + 2Br^-(aq) \rightarrow 2Cl^-(aq) + Br_2(aq)$ (1)

ii Disproportionation (1); Bromine is both oxidized and reduced in the same reaction (1); 0 goes to +1 and −1 (1)

iii +4 (1) and +6 (1)

iv Bromine's oxidation number is reduced from 0 to −1 (1).

b i $SO_2(g) + 2H_2O(aq) + 2e^- \rightarrow SO_4^{2-}(aq) + 4H^+(aq)$ (1)

ii $5SO_2(g) + 4IO_3^-(aq) + 2H^+(aq) \rightarrow 5SO_4^{2-}(aq) + 2H_2O(l) + 2I_2(aq) \times 5$ (1) balance (1)

Total 10 marks

19 a i & ii

Electrons (1); structure (1)

b Three bonding pairs and one non-bonding pair (1); find minimum repulsion arrangement of tetrahedral shape (1); but non-bonding pairs repel more so bond angle is closed down 2.5° from methane (1).

Total 6 marks

20 a i Sodium/potassium hydroxide warmed in ethanol (1)

ii Ammonia in ethanol heated under pressure (1)

iii Heat with sodium/potassium hydroxide (1) in ethanol (1)

b i Bromochlorodifluoromethane (1)

ii Any two from halogenoalkanes absorb heat from fire/prevent oxygen reaching fire/absorb free radicals in combustion/strength of $C-F$ bond makes it inert (2).

iii Halogenoalkanes such as CF_2ClBr can release Cl free radicals; Cl free radicals react with O_3; ozone layer depletes; leading to greater levels of UV exposure; greater risk of skin cancer (Any 3 from above, in context and using correct terminology)
AND CF_3CHF_2 has strong C-F bonds so does not release F radicals (4)

Total 11 marks

21 a i 1.40×10^{-5} mol (1)

ii $I_2(aq) + 2S_2O_3^{2-}(aq)$ (1) $\rightarrow 2I^-(aq) + S_4O_6^{2-}(aq)$ (1)

iii 7.00×10^{-6} (1)

iv $Cl_2(aq) + 2I^-(aq) \rightarrow 2Cl^-(aq) + I_2(aq)$ (1) for LHS; (1) for RHS

v 7.00×10^{-6} (1)

vi 4.97×10^{-4} g (1)

vii 1000 g contains 4.97×10^{-4} g. 1×10^6 g contain 4.97×10^{-1} g (1) which is just less than 0.5 (1).

b Chlorine is toxic at higher concentrations (1); chlorine has an unpleasant taste (1).

c The result is not reliable because the experiment was only carried out once (1); so there is no evidence the experiment is repeatable (1).

Total 14 marks

22 a

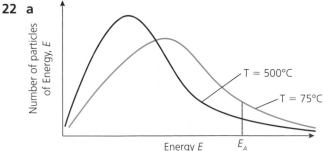

Curves (1); where T_1 has higher peak to left of 750°C peak (1); smaller area under curve above E_A (1).

b By providing an alternative route (1) for a reaction with a lower activation energy (1). So a larger proportion of molecules react at a given temperature (1).

Total 6 marks
Total 47marks

Section C

23 a i To prevent the mixture heating too rapidly (1).

ii Round-bottom/pear shape flask and heat indicated (1); condenser (with drawn jacket) (1); correct water flow (1); the apparatus must not be sealed or be open in the wrong places.

iii Green/blue (1)

b i They have very high boiling temperatures (1).

ii Fractional distillation/distil off water, then ethanoic acid (1).

c i Increased yield (1) as reaction moves to the right because there are fewer gaseous molecules (1).

ii Any two from:
yield of lab process might be low
cost of oxidizing agent
toxicity of oxidizing agent
disposal of Cr^{3+}
control of temperature
lab process has low atom economy
energy cost of separation
lab process is a batch process (2).

d Discussion of four aspects of the process,
Cativa runs at lower pressure hence less energy required
Cativa has 100% atom economy
Methanol in Cativa could be obtained from renewable resources
Cativa produces only one product so less separation needed
Cativa runs at higher temperature so greater energy requirements for heating.
Two additional pieces of information from:
lifecycle cost of catalysts
lifecycle cost of capital equipment
yield of reactions
availability of renewable methanol (4).

Total 17 marks
Total for test = 84 marks